Integrated Landscape Approaches
for Africa's Drylands

A WORLD BANK STUDY

Integrated Landscape Approaches for Africa's Drylands

Erin Gray, Norbert Henninger, Chris Reij, Robert Winterbottom, and Paola Agostini

WORLD BANK GROUP

ISBN (paper): 978-1-4648-0826-5
ISBN (electronic): 978-1-4648-0827-2
DOI: 10.1596/978-1-4648-0826-5

Cover image: Photograph by Andrea Borgarello for World Bank / TerrAfrica. Photo illustration by Luis Liceaga for World Bank / TerrAfrica. © Content creators and World Bank / TerrAfrica. Used with the permission of World Bank; further permission required for reuse.
Cover design: Debra Naylor, Naylor Design, Inc.

Library of Congress Cataloging-in-Publication Data
Names: Gray, Erin, 1982- author.
Title: Integrated landscape approaches for Africa's drylands / Erin Gray, Norbert Henninger, Chris Reij, Robert Winterbottom, and Paola Agostini.
Description: Washington DC : World Bank, [2016]
Identifiers: LCCN 2016018516 | ISBN 9781464808265
Subjects: LCSH: Agricultural landscape management–Africa, Sub-Saharan. | Land use–Africa, Sub-Saharan–Planning. | Natural resources–Africa, Sub-Saharan–Management. | Arid regions–Africa, Sub-Saharan.
Classification: LCC HD108.6 .G73 2016 | DDC 333.73/60967–dc23
LC record available at https://lccn.loc.gov/2016018516

Contents

Boxes

Figures

Photos

Tables

Foreword

Drylands—defined here to include arid, semi-arid, and dry sub-humid zones—are at the core of Africa's development challenge. Drylands make up 43 percent of the region's land surface, account for 75 percent of the area used for agriculture, and are home to 50 percent of the population, including a disproportionate share of the poor. Due to complex factors, the economic, social, political, and environmental vulnerability in Africa's drylands is high and rising, jeopardizing the long-term livelihood prospects for hundreds of millions of people. Climate change, which is expected to increase the frequency and severity of extreme weather events, will exacerbate this challenge.

Most of the people living in the drylands depend on natural resource-based livelihood activities, such as herding and farming. The ability of these activities to provide stable and adequate incomes, however, has been eroding. Rapid population growth has put pressure on a deteriorating resource base and created conditions under which extreme weather events, unexpected spikes in global food and fuel prices, or other exogenous shocks can easily precipitate full-blown humanitarian crises and fuel violent social conflicts. Forced to address urgent short-term needs, many households have resorted to an array of unsustainable natural resource management practices, resulting in severe land degradation, water scarcity, and biodiversity loss.

African governments and the larger development community stand ready to tackle the challenges confronting dryland regions. But while political will is not lacking, important questions remain unanswered about how the task should be addressed. Do dryland environments contain sufficient resources to generate the food, employment, and income needed to support sustainable livelihoods for a fast-growing population? If not, can injections of external resources make up the deficit? Or is the carrying capacity of dryland environments so limited that out-migration should be encouraged as part of a comprehensive strategy to enhance resilience? And given the range of policy options, where should investments be focused, considering that there are many competing priorities?

To answer these questions, the World Bank teamed up with a large coalition of partners to prepare a study designed to contribute to the ongoing dialogue about measures to reduce the vulnerability and enhance the resilience of populations living in the drylands. Based on analysis of current and projected future drivers of vulnerability and resilience, the study identifies promising interventions, quantifies their likely costs and benefits, and describes the policy trade-offs that will need to be addressed when drylands development strategies are devised.

Sustainably developing the drylands and conferring resilience to the people living on them will require addressing a complex web of economic, social, political, and environmental vulnerabilities in Africa's drylands. Good adaptive responses have the potential to generate new and better opportunities for many people, cushion the losses for others, and smooth the transition for all. Implementation of these responses will require effective and visionary leadership at all levels from households to local organizations, national governments, and a coalition of development partners. This book, one of a series of books prepared in support of the main report, is intended to contribute to that effort.

Magda Lovei

Manager, Environment & Natural Resources Global Practice

World Bank Group

Acknowledgments

This book is one of a series of thematic books prepared for the study "Confronting Drought in Africa's Drylands: Opportunities for Enhancing Resilience." The study, part of the Africa Regional Studies Program of the World Bank Group Africa Region Vice Presidency, was a collaborative effort involving contributors from many organizations, working under the guidance of a team made up of staff from the World Bank Group, the United Nations Food and Agriculture Organization (FAO), and the Consultative Group for International Agricultural Research Program on Policies, Institutions, and Markets (CGIAR-PIM). Raffaello Cervigni and Michael Morris (World Bank Group) coordinated the overall study, working under the direction of Magda Lovei (World Bank Group).

This book, entitled "Integrated Landscape Approaches for Africa's Drylands," was prepared by Erin Gray, Norbert Henninger, Chris Reij, and Robert Winterbottom (all of World Resources Institute) and Paola Agostini (World Bank Group).

The book was reviewed by Diji Chandrasekharan Behr, Erick Fernandes, Michael Morris and Raffaello Cervigni (World Bank Group), Mamadou Diakite (TerrAfrica coordinator at the African Union New Partnership for Africa's Development—NEPAD), and a team from FAO. Sara Scherr (EcoAgriculture Partners) reviewed an earlier draft of this book and provided helpful advice to the authors.

Editorial and design services were provided by Luis Liceaga, with support from Madjiguene Seck, Elizabeth Minchew, and Amy Gautam (World Bank Group).

Funding for the book was provided by the TerrAfrica Leverage Fund, the PROFOR Trust Fund, and the World Bank Group Africa Regional Studies Program.

About the Authors

Erin Gray is an Environmental Economist with the World Resource Institute (WRI). Her areas of expertise include natural infrastructure valuation, cost-benefit, cost-efficiency, and multi-criteria analysis, ecosystem service markets and conservation finance, ecosystem-based restoration, and adaptation monitoring and evaluation. Her work concentrates on conducting economic and financial analysis of coastal, agricultural, and forest ecosystems; developing technical and analytical guidance, methods, and tools to support economic valuation and uptake of results by decision makers; designing adaptation monitoring and evaluation guidance for watershed development in India; and supporting efforts to promote water quality trading in the Chesapeake Bay. Erin received her Master of Environmental Management degree from Duke University's Nicholas School of the Environment, concentrating in Environmental Economics and Policy. Erin also holds a Bachelor of Arts degree in both Economics and Environmental Analysis and Policy from Boston University.

Norbert Henninger is a Senior Associate at the World Resources Institute (WRI) working at the intersection of poverty reduction, natural resources management, and governance. His work focuses on creating better information and new tools to formulate and evaluate development cooperation programs, advance green growth strategies, and carry out environmental and social assessments. He has authored or co-authored technical reviews and publications on targeting agricultural research and poverty reduction programs, environmental and agricultural indicators, and integrated assessments of ecosystems and human well-being. He received his MSc in Environmental Sciences from Johns Hopkins University and his MBA from the University of Mannheim, Germany.

Chris Reij is a Sustainable Land Management specialist and a Senior Fellow with the World Resources Institute. He has worked in Africa since 1978. Although he has maintained a focus on the West African Sahel, Chris has been involved in numerous studies and consultancies in other parts of Africa, Asia, the Caribbean and the Pacific. His main fields of research and writing are related to restoration of degraded land in semi-arid regions, farmer innovation in agriculture, long-term trends in agriculture and environment and analysis of successes in agriculture and land management in Africa. He is the facilitator of "African Re-greening

Initiatives," which supports farmers in adapting to climate change and in developing more productive and sustainable farming systems. This initiative was launched to help scale up proven successes in re-greening by individual farmers and communities. It is operational in Burkina Faso and Mali, and is expanding to other African countries.

Robert Winterbottom is a Senior Fellow at the World Resources Institute (WRI) working on the Global Restoration Initiative to restore the productivity of degraded landscapes and scale up improved land and water management. He has broad experience in environmental sciences and natural resource management, including work on desertification control and community-based natural resource management. He is engaged in developing more effective program approaches integrating governance and poverty reduction with environmental management, and in assessing the opportunities for climate change adaptation and investments in REDD+ (Reducing Emissions from Deforestation and Forest Degradation) and landscape restoration. He has a Master of Forest Science (MFS) in Natural Resources Management from the School of Forestry and Environmental Studies of Yale University, and a BA in Geography from Dartmouth College.

Paola Agostini is a Lead Environmental Economist in the World Bank Environment and Natural Resources Global Practice. She is currently the Global Lead for Resilient Landscapes, where she examines projects and programs that try to improve the connectivity of protected areas, forests, agroforestry, rangeland, and agricultural land so as to increase productivity, community resilience, and production of ecosystem services. She holds a PhD in Economics from the University of California, San Diego, and an MSc in Economic and Social Sciences from Università Bocconi, Milan, Italy.

Executive Summary

This book presents emerging findings on the importance of moving beyond single-sector interventions to embrace integrated landscape management that takes into account the health of the ecosystems that support human livelihoods and contribute to the resilience of rural communities in Sub-Saharan African drylands. Integrated landscape management is particularly important for these drylands as people depend on production systems that are frequently disrupted by exogenous shocks such as drought. Households rely on movement over relatively large areas or on diversification of livelihoods. The latter includes the use of multiple plant and animal species, but also migratory labor and transformation of and marketing of agricultural products. This reliance on multiple ecosystems and ecosystem services can become easily unbalanced with an isolated management focus on one sector or one commodity, which in turn may degrade other resources, reduce food security, or increase other risks. Integrated landscape management can help to avoid or minimize the potentially negative impacts of such uncoordinated, sector-specific interventions and capitalize on potential synergies. Preliminary findings from this publication were shared with several audiences to gather input, discuss key results, and refine content. An overview of the book was first presented at a side event during the 2013 United Nations Conference to Combat Desertification Conference of the Parties 11 in Windhoek, Namibia. Additionally, the authors presented interim results at a seminar hosted by the Consultative Group for International Agricultural Research Program on Policies, Institutions and Markets (CGIAR-PIM) at the International Food Policy Research Institute (IFPRI), a practice session at the World Bank Sustainable Development Network Forum, and a PROFOR Advisory Board meeting in 2014.

Challenges

Dryland communities along with their production systems and human livelihood strategies have evolved over hundreds of years in response to an unfavorable climate, enabling both ecosystems and human well-being to recover following droughts, floods, and fires. Over the past decades, however, high human population growth rates, land-use pressures and land degradation, changes in rainfall patterns, greater frequencies and intensities of droughts, conflicts over natural resources, and other natural and anthropogenic drivers have begun to undermine the resilience of many dryland communities in Africa and have contributed to depleted soil fertility and water stress. Local communities are facing a reduced capacity of the land to support them, lowering their resilience to recover from natural shocks.

Although there is an increasing number of positive experiences, efforts to address these challenges in the drylands of Sub-Saharan Africa have too often failed to achieve significant and lasting improvements at scale. Few interventions have been designed to take into account the linkages between upstream farmers and downstream water users. In many cases interventions have disrupted traditional management systems for common pool resources such as wetlands, grazing reserves, and forests.

Single-objective and sectoral development approaches in particular are increasingly seen as inadequate because they may not fully address trade-offs associated with competing land uses and actors, or fall short in incorporating the perspectives of all stakeholders in local communities and in appropriately addressing sources of resource conflict. They may also fail to take into account the biophysical connections and leverage interactions among production systems which are critically important in dryland systems and necessary to generate and sustain both farm-level and landscape-level benefits. For example, trees in agricultural landscapes play a critical role in renewing soil fertility, providing additional sources of fodder for livestock and fuelwood for households, and sustaining cropland productivity, while simultaneously contributing to the diversification and enhanced resilience of farming systems; yet many agricultural and livestock development programs have not taken full account of the key roles of trees in agricultural landscapes. Many development actors across Sub-Saharan Africa are starting to adjust drylands development programs in such a way that they consider multiple objectives and multiple actors across two or more sectors; applying a step-wise, carefully sequenced landscape approach can increase the effectiveness of these programs and capitalize on opportunities to restore resilience in drylands.

Opportunities to Reduce Vulnerability and Increase Resilience

Water scarcity and land degradation are the major biophysical constraints facing drylands and are key threats to economic development and human welfare. Sustainable land management interventions to conserve soil and water, build natural and social capital, and maximize efficiency of water and soil resource use are critical for stabilizing rural production systems. This will also help rebuild household resilience. In many locations, these interventions are the foundation for sustainable agricultural intensification. The following practices have been identified as especially promising for drylands where the need for the widespread adoption of improved land and water management practices to boost productivity is especially acute: agroforestry, farmer-led soil and water conservation techniques, rainwater harvesting, conservation agriculture, and integrated soil fertility management. These measures have been effective in reversing land degradation and in contributing to the sustainable intensification of agriculture and forestry. Rural economies benefit from these practices through higher crop yields, increased supplies of fodder, firewood and other valuable goods, greater income and employment opportunities, a restoration of biodiversity and ecosystem services, and higher climate change resilience. Support for the widespread adoption of these improved land management practices can be a core element of integrated landscape management designed to enhance and diversify production systems and increase household resilience.

Integrated landscape management represents an opportunity to restore dryland areas in Sub-Saharan Africa. The definition adopted for this book is based on that presented by the Landscapes for People, Food and Nature initiative, a collaborative partnership of leading environmental and agricultural nongovernmental organizations (NGOs), UN agencies, and governments.

A long-term collaboration among different groups of land managers and stakeholders is required to achieve the multiple objectives required from the landscape. These typically include agricultural production, provision of ecosystem services (such as water flow regulation and quality, pollination, climate change mitigation and adaptation, cultural values); protection of biodiversity, landscape beauty, identity, and recreation value; and local livelihoods, human health, and well-being. Stakeholders seek to solve shared problems or capitalize on new opportunities that reduce trade-offs and strengthen synergies among different landscape objectives. Because landscapes are coupled with socio-ecological systems, complexity and change are inherent properties that require management.

Integrated landscape management presents an opportunity to scale and leverage land and water management interventions such that the whole is greater than the sum of individual interventions in terms of ecological and economic gains. EcoAgriculture Partners (2013b) identified key actions for operationalizing integrated landscape management to promote successful drylands restoration and community resilience, including: (1) interventions are designed to promote multiple goals and objectives; (2) ecological, social, and economic interactions are managed to reduce negative trade-offs and optimize synergies; (3) roles of local communities are acknowledged; (4) planning and management of interventions is adaptive; and (5) collaborative action and comprehensive stakeholder engagement is encouraged and institutionalized. Based on the EcoAgriculture Partners (2013b) report and research results, this book categorizes these key actions into three broad core components for operationalization, and then provides 10 key principles that can be viewed as a checklist for implementation and operationalization of integrated landscape management (see figure ES.1). The three core components are:

Figure ES.1   Core Components of Integrated Landscape Management

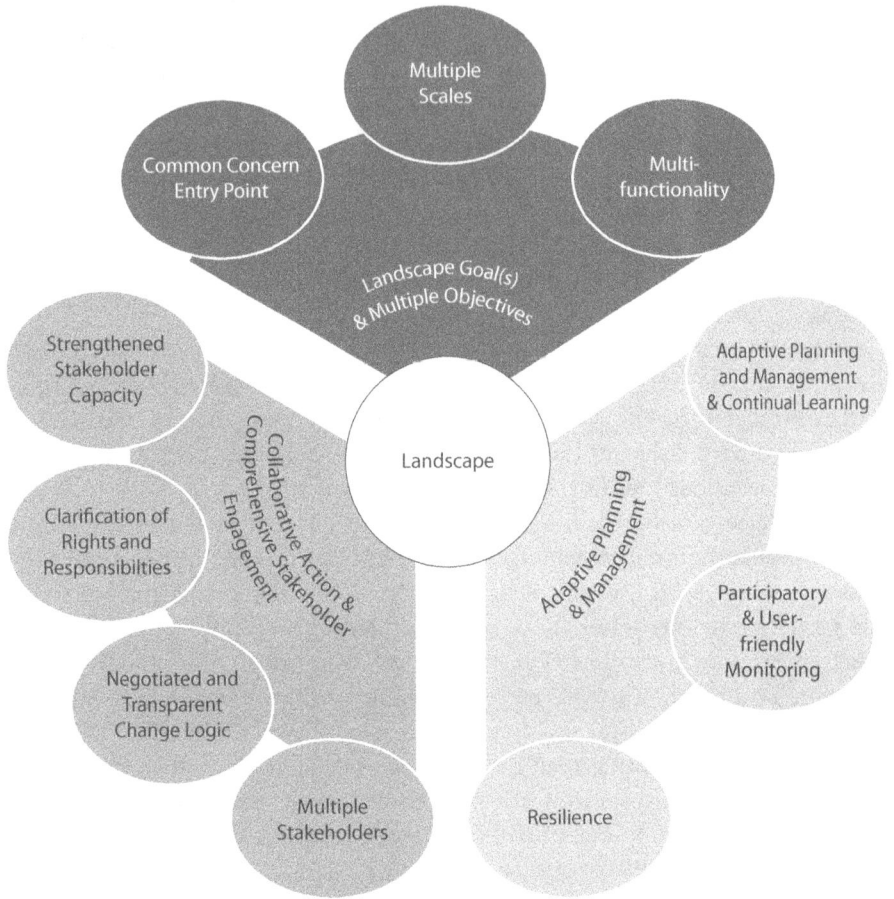

Multiple Scales

Common Concern Entry Point

Multi-functionality

Landscape Goal(s) & Multiple Objectives

Strengthened Stakeholder Capacity

Adaptive Planning and Management & Continual Learning

Collaborative Action & Comprehensive Stakeholder Engagement

Landscape

Adaptive Planning & Management

Clarification of Rights and Responsibilties

Participatory & User-friendly Monitoring

Negotiated and Transparent Change Logic

Multiple Stakeholders

Resilience

Source: Based on Sayer et al. 2013.

Core Component 1: Landscape Goal(s) Encompassing Multiple Objectives at Different Scales

In drylands with mixed land uses and stakeholders, it is important to establish a shared perception of dryland landscapes by identifying and fostering multiple objectives and goals. This promotes common entry points among stakeholders for collaboration in actions that are critical to enhancing resilience. In Sub-Saharan African drylands, goals and objectives generally focus on improving food security and livelihoods diversification. In some higher-producing regions, goals focus on sustainable intensification of production systems. Integrated landscape management must also consider multiple scales at which to implement interventions (for example, farm level and landscape level) as well as temporal and biophysical dimensions, which are especially important in environments with highly variable and unevenly distributed rainfall.

It is important to generate short-term economic returns to incentivize farmer and herder participation, but integrated landscape management also promotes thinking holistically about maximizing ecological gains, for example by improving biophysical connectivity to restore groundwater levels or by providing critical corridors for livestock movement and wildlife habitat. Integrated landscape management must also consider the multi-functionality of landscapes and provide a mechanism to enable local stakeholders to reduce conflicts among different types of specialized resource users (such as herders, farmers, fishers) who differ in their dependencies on a range of ecosystem services.

Core Component 2: Adaptive Planning and Management

Integrated landscape management must seek to understand how land users interact with their environment and key sources of income that can improve welfare. The planning of land use, grazing, and natural resources use under integrated landscape management must recognize ecological, social, and economic interactions among different parts of a landscape, which then can be managed to optimize synergies and reduce negative trade-offs. Integrated landscape management should promote continual learning from outcomes and create opportunities to scale successes and address failures. Adaptive management is also important for understanding the resilience of a landscape or how it responds to shocks such as changes in rainfall and temperature. As climatic and economic risks create uncertainty, adaptive planning and management, whereby stakeholders review at recurring intervals the successes and challenges of current land-use choices, allows all involved to quickly address risks. As such, integrated landscape management requires effective user-friendly participatory monitoring and evaluation systems and feedback mechanisms.

Core Component 3: Collaborative Action and Comprehensive Stakeholder Engagement

Integrated landscape management must recognize that it is critical to identify and acknowledge the roles of local communities and households in resource management. Integrated landscape management must promote community-wide participation and planning in drylands restoration projects and other land-use interventions, collective action for implementing these interventions, and coordination among key stakeholders across scales and sectors. For example, the collaborative actions of farmers on steep slopes in tandem with concerted actions by herders to reduce grazing pressures in critical locations will have greater impact on erosion and sedimentation rates and restoration of the vegetative cover than fragmented or individual efforts alone. Local communities must be incentivized to invest in improved land and water management and share local knowledge and experience.

As stated before, this book adopts 10 key principles (ovals in figure ES.1) based on these three core components for integrated landscape management (Sayer et al. 2013) that are useful to design a process that motivates various stakeholders to pursue a common goal within a landscape, make synergies and trade-offs between objectives more transparent, and establish agreed upon mechanisms to resolve differences between stakeholders.

Skeptics of integrated landscape management may characterize this quest for more integration across multiple sectors and stakeholders and greater emphasis on geographic targeting as nothing new. However, the way integrated landscape management is being proposed conceptually here is different—it builds on lessons learned from previous approaches and places a larger emphasis on building resilience to drivers like climate change and changing market forces. Integrated landscape management provides added value in that it:

• Does not promote a "one-size-fits-all" approach but rather asks stakeholders to consider the local context and include sectors, stakeholders, and social, cultural, and other conditions into account across geographic boundaries that make ecological sense. Integrated landscape management promotes a flexible framework for scaling investments at a landscape level to maximize ecological, economic, and social synergies and minimize negative trade-offs.
• Emphasizes that planning and implementation take into account spatial components that are important to rejuvenate and maintain ecosystem health (for example, hydrological flows, habitat). Integrated landscape management requires that land-use planning and decision making are rethought differently about scale and take into account these spatial components.
• Promotes a combination of bottom-up and top-down principles designed to encourage local community participation, but at the same time dedicated to building appropriate institutional and financial support.
• Promotes an adaptive management approach that tries to build in monitoring and evaluation components to generate long-term data needed to truly understand whether communities are becoming more resilient and increasing their adaptive capacity, and whether landscape-level changes are achieved.

There are many advantages to promoting integrated landscape management in Sub-Saharan Africa's drylands:

• Increased action and investment from stakeholders: Community-driven integrated restoration of small watersheds in Tigray, Ethiopia, for example, has motivated farmers to invest in improved soil and water management practices. Their coordinated efforts have resulted in recharged groundwater tables in valley bottoms, allowing farmers to develop dry season irrigation and cultivate higher value crops.
• Reduced conflict over resources and land uses: Improved coordination among stakeholders can help to clarify rights and responsibilities and improve understanding of landscape goals and objectives and, as such, can lead to reduced conflict over resources and land uses, as in the case of negotiated agreements between farmers and herders to demarcate corridors for livestock movement

that have helped to protect farmers' crops and trees from livestock browsing while safeguarding grazing and water access areas for herders and, in the case of agreements between local communities and firewood and charcoal merchants, to buy wood from locally managed forests and tree farms.

- Economies of scope and scale: Land and water users in a landscape sharing their skills and assets can achieve economies of scale and cost advantages resulting from integrated production. Some landscape interventions can also provide increases in household income associated with the production of two or more products simultaneously.
- Capacity building: As integrated landscape management promotes community participation and collective action, farmers, herders, and other natural resource users learn about new sustainable practices and technologies. Local institutions are empowered to negotiate and adopt rules to improve environmental governance, provide for more equitable benefit sharing, and accelerate the adoption of improved natural resources management practices.
- Resilience at the household and landscape level: Collective action by a large number of households can affect all three dimensions of resilience, depending on the local circumstances: exposure to shocks (for example, households in southern Niger report reduced wind speed at the beginning of the growing season after they increased on-farm tree densities; coping capacity (for example, providing incentives to leave wildlife migration routes unfenced in the Kitengela Plains of Kenya has provided dryland farmers with new income and improved wildlife and tourism benefits for Nairobi National Park); sensitivity to shocks (for example, restoring woodlands and dry season grazing areas through assisted natural tree regeneration in Tanzania's drylands has helped households to diversify livelihood strategies and helped buffer dry season risks for livestock).

Barriers

Several barriers need to be overcome before integrated landscape management can be part of regular policy making and development planning in Africa's drylands:

- Lack of knowledge and awareness about integrated landscape management within national and local governments, private sector, and civil society actors. Landscape-level ideas have yet to percolate down to more national and local actors. Additionally, many integrated landscape management programs lack strong monitoring and evaluation components, especially beyond a household and community scale, making the assessment of landscape-level benefits difficult.
- Institutional barriers that impede addressing complexities at the landscape level. Usually landscapes and the interactions between stakeholders and different land uses are complex. In most cases, there are no simple solutions to complex challenges, and a "one-size-fits-all" approach does not work. A careful assessment of location-specific challenges is required as well as a learning-by-doing approach

and a significant investment in institutional reforms and capacity building. In addition, ways need to be found to take into account sector-specific mandates of different ministries to resolve the challenges of working across sectors.

• Poor availability of and access to location-specific data about land, water, and natural resources use. For many dryland areas, local planners have very limited access to GIS data of land cover, land use, water supply, water extraction, and other natural resources use. This is because data are either nonexistent or are not made publicly available. Without these data it is difficult to develop scenarios of land use that, for example, identify the most promising areas for improved land and water practices to optimize water productivity at landscape scale, or that identify optimal access to riparian areas to increase resilience of pastoral production systems.

• Difficulty in ensuring management of trade-offs and provision of adequate incentives for needed behavior changes and sustainability. In mixed crop production systems, there is a special need to assess trade-offs and synergies between different land uses and users. However, capacity toward this objective among implementing agencies is generally weak.

• Fragmented financing and planning for drylands restoration to optimize land use. Local land-use planning capacity is generally low because of a persistent marginalization of drylands. Failure to address local land-use planning can result in conflict over resources and land areas and other costs.

Recommendations

The ecological and economic evidence gathered in this book shows that integrated landscape management can enhance efforts to invest in tree-based systems and improved livestock management and support productivity increases for rainfed cropping. Integrated landscape management efforts have helped to coordinate the actions of multiple land users and other stakeholders, reduced conflicts, and improved overall governance of water, land, and other resources. Integrated landscape management is thus a useful approach to enhance the intensification of dryland cropping systems and will, in many locations (but not always), result in multiple wins including the following: improved farm productivity, water benefits at farm and landscape level, carbon sequestration, biodiversity and other ecosystem services benefits, and higher climate resilience.

Various policies and related interventions can be used to trigger and accelerate the scaling up of these benefits through integrated landscape management across Sub-Saharan African drylands to restore and increase household and ecological resilience. Policies are needed to develop the framework conditions necessary to both initiate new programs and modify and scale up existing restoration and resilience efforts in Sub-Saharan African drylands. Table ES.1 highlights policy options, covering six broad intervention areas that can be explored and applied to the local context to address the important challenges above and advance integrated landscape management.

Table ES.1 Major Intervention Areas and Associated Policy Options to Advance Integrated Landscape Management

Intervention area	Example policy options
1. Clarify land rights and responsibilities	• Decentralize policies for natural resources management and provide more authority to community resource organizations to empower them to make decisions. • Reform land-use planning and land tenure policies to increase community ownership over resources and create greater security of access to and use of natural resources.
2. Encourage multi-stakeholder involvement and collective action	• Create or invest in incentive schemes to compensate losers and encourage their participation (for example, payment for ecosystem services schemes) in integrated landscape management initiatives. • Encourage policies that support framework conditions for collaborative action and comprehensive stakeholder engagement such as formation and legal standing of common interest groups.
3. Overcome institutional barriers to integrated landscape management	• Conduct a thorough review of current drylands restoration programs and policies to: identify barriers to scaling up restoration success stories and implementing integrated landscape management principles; identify gaps in staffing and policies needed to promote integrated landscape management; and better target funding for drylands restoration activities and programs. • Create a common set of guidelines for drylands restoration that incorporates principles of good practice for integrated landscape management with endorsement from relevant government agencies to show solidarity in promoting integrated landscape management.
4. Create conditions for adaptive planning and management	• Encourage spatial aspects of planning in local development planning and strengthen participatory land-use planning policies where they do not exist. • Create incentives for more coordinated and systematic planning and for linking government budgets and planning.
5. Create mechanisms and supporting policies for sustainable and long-term financing of integrated landscape management	• Work with Ministries of Finance, donors, and other financiers to promote the consolidation of financing for mainstreaming improved land and water management. Increase financing for integrated landscape management in targeted dryland landscapes where rural households are particularly vulnerable and where there are significant opportunities for scaling up agroforestry and other improved land and water management practices. • Create incentives for and reduce perceived risks of integrated landscape management to encourage public and private investments through risk reduction guarantees and other risk reduction mechanisms.
6. Invest in a solid evidence base and knowledge-sharing platforms for integrated landscape management	• Invest in research institutes, academic extension programs, and NGOs to establish and track important restoration indicators, create knowledge-sharing platforms, and establish monitoring and evaluation systems that support long-term analysis and adaptive management (and link to existing drought monitoring and early warning systems). • Identify champions and leaders of integrated landscape management who can play a critical role in raising awareness and promoting this approach and represent different cultural and resource groups and sectors. These champions can be assisted by investing in opportunities to take the lead in documenting integrated landscape management successes.

Note: NGOs = nongovernmental organizations.

These general policy recommendations can be further tailored depending on the agro-ecological zone. For areas with severe degradation, it might be necessary to invest first in sustainable land and water management practices covering social, technical, and environmental aspects to regenerate ecosystem health (for example, increased livestock mobility, livestock exclosures, grazing bans, rainwater harvesting infrastructure, education and skills development for primary stakeholders, social safety net programs to provide food and cash for villagers' implementation of interventions, etc.). Such programs should take advantage of low-cost options (such as catalyzing and building upon innovative farmer-led restoration efforts) and be tailored to the local context which is based not only on who is involved, but also the farming system classification. Table ES.2 highlights examples, which need to be further developed based on local conditions, for three important dryland farming systems.

Table ES.2 Integrated Landscape Management Programs in Three Dryland Farming Systems

System	*Integrated landscape management programs can bolster resilience and scale up sustainable production systems by:*
Mixed crop farming systems (dry sub-humid)	• Reducing conflicts and avoiding negative externalities of intensification • Setting up institutions for integrated land-use planning (that document and take into account existing rights including access to common pool resources and riparian areas) and mechanisms for conflict resolution • Supporting sustainable crop and livestock intensification • Exploring opportunities to take advantage of landscape structure (mosaic of natural and managed ecosystems) to enhance biological control, pest management, pollination, or other ecosystem services • Safeguarding upstream water supplies and reducing negative externalities on other water and natural resource users downstream
Agro-pastoral systems	• Reducing risks related to water shortages and land degradation • Supporting regeneration of dry forests and woodlands through assisted natural regeneration of trees and increasing density of trees on farms through farmer-managed natural regeneration • Safeguarding dry season grazing reserves (for example, wetlands, dry forests, and woodlands) and encouraging planned grazing management • Developing water infrastructure that is aligned with sustainable forage management • Diversifying income sources and increasing coping capacity • Establishing rewards or payments for biodiversity conservation, wildlife corridors and, in steep areas, watershed protection • Reducing conflicts and avoiding decoupling of other resource users • Establishing corridors for livestock movement to protect farmers' crops and trees, designating grazing and water access areas to ensure resilience of pastoral production systems, and setting up a dispute resolution mechanism • Establishing corridors for wildlife to access water and biomass during droughts
Pastoral systems	• Reducing risks related to water shortages and land degradation • Supporting regeneration of pastoral landscapes through assisted natural regeneration of trees and shrubs with the help of exclosures and community-based natural resource management institutions (including customary pastoral institutions) • Safeguarding dry season grazing reserves (including wetlands) and access to and use of natural resources • Supporting pastoral mobility • Developing water infrastructure that is aligned with forage availability and grazing patterns to avoid risk of degradation • Encouraging grazing management that improves soil cover, increases water infiltration and retention, and improves plant diversity and biomass • Diversifying income sources and increasing coping capacity • Establishing rewards or payments for biodiversity conservation

Key Take-Home Messages

Restoring degraded drylands by increasing support for integrated landscape man-agement as a framework for scaling up improved land and water management practices in targeted areas can enhance the resilience of the most vulnerable herders and farmers and become the entry point for extending integrated land-scape management across Africa's drylands. Increased investment in integrated landscape management programs, which support coordination and long-term collaboration among different groups of land managers and stakeholders within dryland landscapes, can enhance and safeguard these restoration efforts, lower risks related to water shortages and land degradation, diversify income sources, support sustainable intensification, and reduce conflicts.

Abbreviations

AIDS	acquired immune deficiency syndrome
CAADP	Comprehensive Africa Agriculture Development Programme
CBNRM	Community-Based Natural Resources Management
CBPWD	Community-Based Participatory Watershed Development
CFAs	Community Forests Associations
CGIAR-PIM	Consultative Group for International Agricultural Research Program on Policies, Institutions, and Markets
CIGs	Common Interest Groups
CSA	climate-smart agriculture
ESI	environmental service index
EU	European Union
FAO	Food and Agriculture Organization
FDA	Focal Development Areas
FMNR	farmer-managed natural regeneration
FSP	Food Security Program
GDP	gross domestic product
GEF	Global Environment Facility
HIV	human immunodeficiency virus
IFAD	International Fund for Agriculture Development
IFPRI	International Food Policy Research Institute
ISRIC	International Soil Reference and Information Centre
IUCN	International Union for the Conservation of Nature
LLPPA	Local-Level Participatory Planning Approach
MKEPP	Mount Kenya East Pilot Program
MoA	Ministry of Agriculture
MoARD	Ministry of Agriculture and Rural Development
NEPAD	New Partnership for Africa's Development
NGO	nongovernmental organization
OECD	Organisation for Economic Co-operation and Development

PES	payments for ecosystem services
PROFOR	Program on Forests
PSNP	Productive Safety Net Program
RISEMP	Regional Integrated Silvopastoral Ecosystem Management Project
SLM	sustainable land management
SLMP	Sustainable Land Management Program
UTaNRMP	Upper Tana Catchment Natural Resources Management Project
WFP	World Food Programme
WRI	World Resources Institute
WRUA	Water Resource Users' Association

CHAPTER 1

Objective, Audience, and Key Questions

The objective of this book is to explore how integrated landscape management can reduce the vulnerability and enhance the resilience of populations living in the drylands of Sub-Saharan Africa. To do this, the book first provides a synthesis of existing knowledge on how integrated landscape management is defined and implemented. Next, it documents key issues related to stakeholder involvement and potential policy instruments to advance integrated landscape management, as well as unique and additional ecological and economic cost and benefits. These concepts are discussed through three detailed case studies of integrated landscape management initiatives in Sub-Saharan Africa. This review then provides the foundation to formulate policy-related recommendations on how to advance integrated landscape management in Sub-Saharan African drylands to reduce peoples' vulnerability and enhance their resilience. This book builds on work conducted by the Landscapes for People, Food and Nature initiative and a wealth of other literature and draws from a broad variety of interventions such as climate-resilient food production programs and integrated watershed management efforts.

The primary audience of this book will be the international community including government officials and development cooperation experts designing and implementing sustainable development programs in the drylands, with a focus on Africa. An associated shorter condensed summary will target senior policy makers in Sub-Saharan Africa and development cooperation and private sector leaders concerned with the sustainable economic development of Africa's drylands.

The findings of this book are expected to help these audiences answer the following questions:

1. What is integrated landscape management in theory and practice and what core principles and best practices have emerged in recent years?
2. What are the roles of different stakeholders within a landscape including private and public actors and how can private farm and farm-level interests be balanced with wider public interests of a landscape approach?

3. What is the economic and ecological evidence of the added value of integrated landscape management in addressing the major sustainable development challenges for drylands and contributing to increasing resilience and reducing vulnerability in the drylands of Sub-Saharan Africa?
4. What has been the experience with integrated landscape management (or with some major dimensions and elements of integrated landscape management) to reduce the vulnerability and enhance the resilience of populations living in Africa's drylands?
5. What are some important lessons learned from the experience gained in developing and implementing integrated landscape management?
6. What policy changes and other interventions are needed to overcome important barriers and advance integrated landscape management in Africa so that it can promote more sustainable landscape management to enhance resilience and reduce vulnerability in Africa?

The book answers the questions outlined above and consists of the following additional chapters:

2. Conceptual Framework
3. Role of Public and Private Stakeholders to Implement Integrated Landscape Management
4. Economic and Ecological Evidence for Integrated Landscape Management
5. Case Studies for Integrated Landscape Management in African Drylands
6. Recommended Policies and Other Interventions Needed to Advance Integrated Landscape Management and Enhance Resilience in Drylands

CHAPTER 2

Conceptual Framework

Introduction

This introductory section describes the recent emergence of the landscape concept and integrated landscape management. While a more detailed definition is provided later in this section, integrated landscape management is the long-term coordination and collaboration of resource users and land managers to restore the productivity and resilience of rural landscapes to achieve multiple ecological and economic benefits from a geographic area. This section highlights some of the key proponents of this new approach and compares it with other integrated geographic approaches that have been proposed or applied in the past. It concludes with a summary of the challenges that need to be addressed to encourage a more widespread use within national policy making and international development cooperation.

Stronger Voices in Support of a Landscape Approach

A brief scan of background papers, blog posts, and presentations leading to and following the RIO+20 meeting and other international meetings (for example, the United Nations Convention to Combat Desertification Conference of the Parties [UNCCD COP] 11 in Namibia in September 2013 and the Global Landscapes Forum in Poland in November 2013) reveals an emergence of the landscape concept to address economic development challenges. International institutions such as the World Bank and members of the international agriculture research community are encouraging a more integrated geographical and socio-economic approach to manage land, water, and forest resources to promote multiple objectives including food security, reduce negative environmental impacts, and identify new pathways to transition to a green economy.

For international financial institutions such as the World Bank, one major reason to promote a landscape approach is to design programs that address better the interconnected challenges related to poverty reduction and food production in the face of climate change, water scarcity, land degradation, and deforestation (for example, see the World Bank's "Landscape Approaches in Sustainable Development" or "What Is a Landscape Approach and Why Is It Necessary?"). A second motivation relates to the realization that investments in natural capital to restore soils or tree cover can make ecosystems, and more specifically farming systems, more robust to withstand external shocks such as climate change (for example, the Program on Forests (PROFOR), a multi-donor partnership housed at the Bank, is encouraging investment in trees and landscape restoration in Africa; see "PROFOR— Sustaining Forests for All"). A landscape approach is also needed to manage many tree and forest resources that exist outside of discrete, remaining large blocks of natural forest (Dewees 2013).

International research organizations such as the Center for International Forestry Research (CIFOR), World Agroforestry Centre (ICRAF), and selected Consultative Group for International Agricultural Research (CGIAR) consortium research programs are seeking to address issues related to climate change, poverty, and food security in a more integrated manner as well. These institutions perceive a landscape approach as a useful bridge that overcomes the divide between the forest and agriculture sectors. Researchers are beginning to focus on research questions that examine, for example, how a landscape can provide a better mix of on- and off-farm benefits and improve the delivery of ecosystem services such as water purification, water retention, soil fertility, and carbon sequestration. The findings from such research can help policy makers determine more optimal land-use patterns within a geographic area that result in a more balanced distribution of productivity gains, livelihood opportunities, and negative environmental impacts (for example, see Holmgren "Landscapes for Sustainable Development"; or discussed in context with new Sustainable Development Goals).

This increased interest to understand and promote investments that account for the relationship between natural capital and its ecosystem services, especially within agricultural landscapes, is at the core of new platforms for cross-sectoral dialogue, learning, and action. The Landscapes for People, Food and Nature initiative, for example, is a collaborative effort of international and national institutions to understand and support integrated agricultural landscape approaches that simultaneously meet goals for food production, ecosystem health, and human well-being.

Development Approaches Are Evolving toward Greater Integration of Multiple Sectors and Scales

Over the past few decades, various integrated geographic approaches have been proposed and applied in development programs such as: community-based natural resources management, integrated natural resources management, sustainable land (and water) management, and climate-smart agriculture. Table 2.1 provides some examples of early critiques of these approaches and how they have changed over time.

Table 2.1 Sample of Development Approaches Used in Africa

Development approach	Lead proponents (examples)	Goals and objectives	Stakeholder engagement	Critiques, limitations, and difference from integrated landscape management	Recent developments
Integrated Conservation and Development Projects (ICDPs) *First introduced in the 1980s by NGOs like the World Wildlife Fund.*	• International conservation organizations	• Aspires to reconcile biodiversity and conservation issues with rural development issues (address livelihood needs of local communities). ICDPs sought to address biodiversity conservation using socio-economic investment tools.	• Initial projects had limited level of community participation.	• Focused narrowly on biodiversity conservation and neglected the agricultural and other sectors. • Implementing organizations critiqued for not understanding the social and economic dynamics of the areas they were applying this approach. • Implementation based on social units that did not match with social organizational structures.	• Efforts have evolved into more integrated approaches where agriculture is an important land use. • Additionally, there has been a trend towards integrating adaptive management into projects. • More recent CBNRM projects have incorporated a landscape perspective such as USAID CARPE project in Central Africa. • More recent CBNRM projects have emphasized governance and power issues and the linkages to natural resources and livelihoods (for example, USAID's Nature-Wealth-Power approach).
Community-Based Natural Resources Management (CBNRM) *CBNRM began gaining traction in the 1970s and has been applied widely around the world.*	• USAID • Environmental NGOs (for example, World Wildlife Fund) • United Nations	• This approach is rooted in the belief that natural resource users (that is, local communities) are best suited to conserve natural resources. CBNRM thus aspires to empower local communities by improving their access and use of natural resources.	• Local community level.	• Integrated conservation-based enterprise development is an important component. • Initially, focused too narrowly on community institutions and not on governance. • Faced difficulty in reconciling objectives of socio-economic development, biodiversity protection, and sustainable resource use.	

table continues next page

5

Table 2.1 Sample of Development Approaches Used in Africa *(continued)*

Development approach	Lead proponents (examples)	Goals and objectives	Stakeholder engagement	Critiques, limitations, and difference from integrated landscape management	Recent developments
Farming Systems Approach (FSA) ("Gestion de terroirs" in West Africa) *Popular in the 1980s–90s*	• USAID • World Bank	• More holistic approach to agricultural development.	• Often participatory—promoted involvement of farmers in decision-making at the village-level.	• Narrowly focused on agriculture. • High start-up costs. • Slow pace of participatory procedures (without producing immediate benefits to farmers). • Projects were not linked to administrative structures (institutional vacuum). • Did not properly address ecosystem management issues, and the link between ecosystems and food security and livelihoods. • Early efforts paid little attention to policy and infrastructure and incorporation of technology that met farmers' needs.	• Ongoing decentralization process in various countries has strengthened local institutions and planning efforts. This has resulted in better institutional linkages for *gestion de terroirs* projects and the respective planning and implementation processes can be better aligned.
Integrated Natural Resources Management (INMRM) *1998*	• CGIAR	• Aspires to align food production, livelihood security, and ecosystem management and promotes a multi-sectoral approach.	• Promotes a multi-stakeholder approach	• INRM concept too complex to translate into action • Expensive to implement through projects	• New technology and new commitments by donors in landscape and system approaches for research.
Integrated Watershed-Based Approaches (for example, Participatory Integrated Water Resources Management; Community-based Participatory Watershed Development; Integrated Water Resources Management (IWRM; Watershed Development)	• National governments (for example, India and Ethiopia)	• Understand and optimize water resources across social, economic, and ecological needs.	• Recent successful efforts are including bottom-up approaches building on local level participatory planning approaches for integrated natural resources management.	• Early integrated watershed projects were top down and too large (for example, 300–400 km²), lacked efficient community participation, created limited sense of responsibility over assets created, and used unmanageable planning unit.	• Over the past three decades, integrated watershed projects have moved beyond focusing on water to better recognize the inter-linkages with agriculture and livestock. Additionally, many projects now promote a community-led or bottom-up approach.

table continues next page

Table 2.1 Sample of Development Approaches Used in Africa (continued)

Development approach	Lead proponents (examples)	Goals and objectives	Stakeholder engagement	Critiques, limitations, and difference from integrated landscape management	Recent developments
Integrated watershed approaches began gaining popularity in the 1980s in semi-arid and arid regions and are still used today.			• Ideal planning units 50 km² with investments in small-scale community infrastructure	• Prioritized water quantity and quality; did not prioritize other ecosystem services that relate to water quantity and quality provision. • Many projects promoted a ridge-to-valley approach and did not consider neighboring watersheds.	
Sustainable land management (SLM) Sustainable land and water management (SLWM) *Began in the 1990s.*	• Food and Agriculture Organization of the United Nations (FAO) • UN Earth Summit 1992 • TerrAfrica • The Comprehensive Africa Agriculture Development Programme (CAADP) • The World Bank • UNDP-GEF • National governments (for example, Ethiopia)	• SLM and SLWM aspire to optimize the use of land resources, including soils, water, animals and plants for the production of goods to meet needs, while simultaneously ensuring the long-term productivity of these resources.	• SLM and SLWM were initially more top-down approaches that lacked community/local participation.	• Entry points are farmers' and other resource users' practices (with focus on technologies and approaches), hence spatial aspects and associated landscape levels are secondary effects (for example, aggregation of individual efforts). • A landscape objective (linked to a geographic area) is not the initial entry point; hence stakeholders do not systematically seek to solve shared problems or capitalize on new opportunities that reduce trade-offs and strengthen synergies among different land-use options. • No systematic long-term effort to identify additional synergistic cost-savings and concurrent impacts linked to landscape scale and stakeholder coordination.	• More recent definitions (SLWM) have expanded original definitions and acknowledge coordinated efforts at multiple scales (for example, farm, watershed, landscape, and global). • CAADP definition seeks to maximize the economic and social benefits from the land while maintaining or enhancing ecological support functions of the land resources. • Most recent efforts incorporate good practice principles from integrated landscape management creating considerable overlap between these approaches.

table continues next page

Table 2.1 Sample of Development Approaches Used in Africa *(continued)*

Development approach	Lead proponents (examples)	Goals and objectives	Stakeholder engagement	Critiques, limitations, and difference from integrated landscape management	Recent developments
Climate-Smart Agriculture (CSA) *Introduced in 2009 by the Food and Agriculture Organization of the United Nations (FAO), and then at the First Global Conference on Agriculture, Food Security and Climate Change at the Hague*	• Environmental NGOs • The World Bank • FAO • CGIAR • United Nations	• CSA aims to sustainably increase agricultural productivity (for food and income security) while building resilience and reducing greenhouse gas emissions from the agricultural sector. • Lower greenhouse gas emissions and/or emissions intensity from the agricultural, food, and forestry sectors. • Sustainably increase agricultural production.	• Often participatory—but is not a "must"	• Focuses too narrowly on the agricultural sector and can fail to recognize linkages with other ecosystem services and broader social dynamics. • Concept is too open whereby a wide range of agricultural interventions could be classified as "climate-smart." • Perception that CSA could potentially marginalize smallholder farmers (if the focus is on maximizing mitigation benefits).	• Efforts are being made to apply climate-smart agriculture cross-sectorally across cropland, livestock, forest and fishery sectors and work at landscape scale (design climate-smart landscapes). • Progress in forging a consensus on key indicators for climate-smart agriculture and in developing monitoring systems to assess the relative effectiveness of various CSA practices. • Momentum building for regional and international alliances to mobilize support for scaling up CSA practices.

Sources: African Union and NEPAD 2009; GIZ 2011; Hughs and Flintan 2011; Kellert et al. 2000; Lakew et al. 2005; Milder et al. 2014; Neufeldt et al. 2013; Sayer et al. 2013; USAID 2012; Verhagen et al. 2014.

Note: NGOs = nongovernmental organizations; CARPE = Central Africa Regional Program for the Environment; CGIAR = Consultative Group for International Agricultural Research; INRM = Integrated Natural Resource Management; UNDP-GEF = United Nations Development Programme–Global Environment Facility; USAID = U.S. Agency for International Development.

The approaches in table 2.1 have been learned from failures of applying other, mainly sectoral strategies. These shifts in scope and implementation are beginning to coalesce along a common set of principles, all of which are becoming part of the envisioned landscape approach. Major trends include:

- *Widening of a single-sector approach*. Many of the examples started out in a single sector (for example, agriculture, biodiversity conservation), then widened their scope to include other sectors and multiple objectives, because of the realization that the dependence and impact on ecosystem services mattered or that livelihoods and poverty reduction objectives must be addressed to ensure uptake and scaling of the interventions. Additionally, approaches have evolved to encourage a more systematic assessment of trade-offs and synergies related to the involved sectors.
- *Adjusting spatial approaches to practical units or scale and full participation of stakeholders*. In many instances, some of these practices have become geographically more explicit, targeting their interventions to specific habitats or buffer zones surrounding protected areas or encouraging interventions within a certain agro-ecological zone or catchment area. In addition, numerous examples started out with a strong spatial component (for example, regional master plan, large water management plan), often using a top-down approach. Over time these interventions have learned from failure.
- *Encouraging comprehensive stakeholder participation, especially of beneficiaries*. Many approaches that were heavily top-down have altered their strategy to become more participatory and have begun working at scales that are more congruent with the problem to be solved, more appropriate to build and sustain assets within the geographic area (for example, building and maintaining sustainable land and water management infrastructure investments), and more optimal in size for successful collective action (for example, watershed development and sustainable land management).
- *Promoting large-scale restoration of a whole landscape or ecosystem*. Examples include the restoration of the Loess Plateau in China (World Bank Institute 2010), investment in sustainable land management in the central plateau of Burkina Faso (Reij, Tappan, and Belemvire 2005), or the rehabilitation of a degraded ecosystem upon which a large geographic area depends such as the Florida Everglades (Doyle and Drew 2008). The effort generally starts out with bringing together all major land and water users (for example, cities, rural communities, agricultural producers, tourism operators, manufacturing facilities, national parks) to decide about the optimal mix of land cover and land use and where to invest in intensification of agricultural production and restoration of ecosystems. Applications that intentionally use a multistakeholder and multi-objective effort to achieve landscape-level outcomes—a plan and implementation to change land cover and land-use patterns across the entire geographic area—are relatively new (Milder et al. 2014).

What Is Driving the Interest in a Landscape Approach?

What has inspired change for these previous approaches? The literature recognizes the following drivers as being behind the shift toward a landscape approach:

- *Realization that geography matters.* The management decisions by farmers, herders, municipalities, companies, conservationists, and other land and water users in a landscape affect each other. The degree to which these individual decisions will result in synergies or trade-offs and determine the resilience of communities are greatly dependent on location-specific factors such as agro-ecological zone, upstream or downstream position within a watershed, water availability, natural resource conditions, or ecosystem health. Proponents of more coordinated land and water management within a geographic area are recognizing that it is in rural landscapes where the linked challenges of poverty reduction, food, water and energy security, ecosystem management, and climate change must be solved (Milder et al. 2014).
- *Success of integrated, bottom-up approaches.* Evidence from programs using participatory integrated watershed management approaches (for example, Kerr 2002; Joshi et al. 2005; WFP 2005; Kale, Manekar, and Porey 2012) and efforts to establish payments for ecosystem services within a landscape or watershed (for example, Lipper et al. 2009; Pagiola 2008; Pagiola, Rios, and Arcenas 2008) show that such integrated and geographically targeted approaches can successfully deliver a more optimal supply and distribution of ecosystem services and provide both private benefits on farms and public benefits within the landscape and watershed.
- *Failure of separate sectoral approaches to halt ecosystem degradation globally.* Narrowly focused sectoral interventions are more frequently seen as unsustainable and with a high risk of generating negative externalities and undermining the provision of critically important ecosystem services at the landscape level (Millennium Ecosystem Assessment 2005; Winterbottom et al. 2013).
- *Renewed interest in investing in agriculture and rural areas.* Over the past decade, various institutions are increasingly promoting investments in agriculture to boost economic development and improve food security including New Partnership for Africa's Development (NEPAD; for example, The Comprehensive Africa Agriculture Development Programme—CAADP), the World Bank (for example, *World Development Report 2008: Agriculture for Development*; World Bank Group Agriculture Action Plan: FY2010–12; and Global Agriculture and Food Security Program—GASFP), or the Gates Foundation (for example, Gates Foundation Agricultural Development Strategy).
- *Growing awareness about potential synergies and trade-offs among different sectors.* Discussions on the nexus of water, energy, and food (for example, NEXUS, the Water, Energy and Food Security Resource Platform) seek to find new approaches to address food security, energy production, economic development, and climate change issues simultaneously. Similarly, some private sector actors are committed to more integrated and cross-sectoral

approaches, especially for agricultural investments (for example, World Economic Forum 2010).

- *Need to take more account of climate change in development strategies and planning.* Increasingly, all investments need to address issues of adaptation, mitigation, and resilience. Proponents of "climate-smart agriculture," for example, demand a more comprehensive accounting of the environmental, health, and social costs of unsustainable agricultural practices, and aim for greater success in achieving a "triple win" of increasing productivity along with adaptation and resilience, and mitigating greenhouse gas emissions while contributing to the achievement of national food security and development goals. They are also encouraging investments that take into consideration the on- and off-farm linkages of agricultural production within a watershed, the amount of fossil fuel inputs, and the level of carbon stored (for example, Kenya Agriculture Carbon Project, see World Bank 2014b; Shames, Wekesa, and Wachiye 2012) and the challenges of implementing such a project (Sharma 2012).

- *Shift in thinking toward investing in natural capital or infrastructure.* Awareness is increasing that healthy ecosystems and investing in "natural capital" are essential for sustaining long-term economic growth and that "working with nature" and "not against nature" can lead to more sustainable results. This shift in thinking can be taken into consideration by decision makers of possible investments to secure water supplies and to manage watersheds as they explore the cost and benefits (for example, Talberth et al. 2012) and long-term sustainability of investing in "gray" (for example, dams, reservoirs, water treatment facilities) and "green" infrastructure (for example, restoration of wetlands, water harvesting on farmland). Comparable pro-nature thinking can also be observed when scanning the literature on climate change adaptation, for example studies that encourage greater investments in ecosystem-based adaptation and in programs aimed at a restoring resilience, especially among vulnerable populations (for example, World Bank 2009a).

- *New information and technology to analyze and communicate land-use issues in a landscape.* Over the past decade the cost of remote sensing products providing up-to-date land cover and land-use information, easily accessible mapping programs, and increased knowledge and modeling tools to simulate water flows and ecosystem delivery are all making it more practical to examine the complex environmental, social, and economic interactions within a geographic area. A landscape approach becomes more feasible with these new tools which allow analysts and stakeholders to explore different land use scenarios, assign associated costs and benefits based on stakeholder input, and communicate policy options to decision makers (for example, Goldstein et al. 2012; Goldstein et al. 2010).

Barriers to the Implementation of a Landscape Approach

The quick review of interventions that seek to achieve landscape-scale benefits for food security, ecosystem health, improved watershed management, climate resilience, or other development objectives makes it clear that the entry points and purposes of these integrated geographic approaches vary greatly. A number of barriers need to be overcome before a landscape approach can be part of regular policy making and development planning or become a new paradigm for international development cooperation:

- *Lack of knowledge and awareness about integrated landscape management within national and local governments, private sector, and civil society actors.* Landscape-level ideas have yet to percolate down to more national and local actors. The continental review of integrated landscape initiatives by the Landscapes for People, Food and Nature initiative found, for example, that within Africa, landscape-level ideas had stayed mostly within the realm of international organizations (Milder et al. 2014).
- *Lack of clarity and imprecise use of terminology, and fragmentary understanding of landscape approaches.* Landscape and integrated approaches have been promoted by various disciplines and sectors. Analysts and decision makers are confronted with too many concepts and definitions spread across disciplines. There is a large body of literature on landscape approaches and ecosystem approaches, often used interchangeably (Sayer et al. 2013). Experiences with landscape approaches are spread widely across several academic fields and communities of practice. This has been the impetus for the Landscapes for People, Food and Nature initiative to carry out a systematic compilation of case studies for Africa, Asia, and Latin America to understand the design, objectives, and results of these initiatives (Milder et al. 2014; Estrada-Carmona et al. 2014).
- *Institutional barriers that impede addressing complexities at the landscape level.* Usually landscapes and the interactions between stakeholders and different land uses are complex. In most cases, there are no simple solutions to complex challenges, and a "one-size-fits-all" approach does not work (Sayer et al. 2013). A careful assessment of location-specific challenges is required as well as a learning-by-doing approach and a significant investment in institutional reforms and capacity building. In addition, ways need to be found to take into account sector-specific mandates of different ministries to resolve the challenges of working across sectors.
- *Poor availability of and access to location-specific data about land, water, and natural resource use, and limited local planning capacity to optimize land use.* For many dryland areas, local planners have very limited access to GIS data of land cover, land use, water supply, water abstraction, and natural resource use. Without these data it is difficult to develop scenarios of land use that, for example, identify the most promising areas for improved land and water practices to optimize water productivity at landscape scale. In addition, local planning capacity is generally low because of a persistent marginalization of drylands. These constraints are especially acute in the mixed production systems of dry sub-humid landscapes, where there is a special need to assess trade-offs and synergies between different land uses.

- *Difficulty in ensuring management of trade-offs and sustainability.* Planning and organization of interventions at the scale of a targeted landscape does not automatically lead to sustainability. Success with a landscape approach requires that trade-offs are well managed by all stakeholders involved. A case in point is the experience with integrated water resources management. Numerous countries have established integrated water resources management plans at various scales, but have not always succeeded in changing unsustainable water use patterns or improving water quality trends (see discussion on the pros and cons of integrated water resources management by Thalmeinerova and Downey 2013 and Giordano and Shah 2014).

Defining Integrated Landscape Management

This section introduces existing definitions of a landscape and defines integrated landscape management and important principles associated with this approach. It is followed by a discussion of benefits and risks associated with integrated landscape management. It also contrasts sectoral and landscape interventions and highlights the specific relevance of integrated landscape management to drylands.

What Is a Landscape?

Various institutions and development programs have defined a landscape in relation to their specific interventions and program goals:

- TerrAfrica in working to scale up "sustainable land management" across sectors.
- The World Bank as part of its work to support agriculture and rural development.
- EcoAgriculture Partners and other co-organizers (including World Resource Institute [WRI], ICRAF, Food and Agriculture Organization of the United Nations [FAO], United Nations Environment Programme [UNEP], International Fund for Agriculture Development [IFAD], Conservation International [CI], and others) of the Landscapes for People, Food and Nature (LPFN) initiative providing analysis of integrated landscape approaches.
- Global Partnership on Forest Landscape Restoration (GPFLR) in its work to restore "forest landscapes" and achieve the targets of the "Bonn Challenge."
- CIFOR in the context of its focus on advancing "sentinel landscapes for sustainable development."
- ICRAF in its efforts to promote land care and to encourage agroforestry, the EverGreen Agriculture Partnership, and landscape restoration.
- International Union for the Conservation of Nature (IUCN) as part of its Livelihoods and Landscapes program.
- PROFOR, FAO, IUCN, WRI, and others committed to building resilient landscapes for improved food security through forestry and tree-based ecosystem approaches, climate-smart agriculture, landscape restoration, re-greening, and related approaches.

- The partners of the Natural Capital Project are applying a landscape approach to optimize ecosystem services management and biodiversity conservation (box 2.1 provides a sample of two landscape definitions).

This publication uses a definition proposed by the Landscapes for People, Food, and Nature initiative (EcoAgriculture Partners 2013b):

> A "landscape" is a socio-ecological system that consists of a mosaic of natural and/ or human-modified ecosystems, with a characteristic configuration of topography, vegetation, land use, and settlements that is influenced by the ecological, historical, economic, and cultural processes and activities of the area. The mix of land cover and use types (landscape composition) usually includes agricultural lands, native vegetation, and human dwellings, villages and/or urban areas. The spatial arrangement of different land uses and cover types (landscape structure) and the norms and modalities of its governance contribute to the character of a landscape.

Since the management objectives for a landscape can vary greatly, its size and boundaries vary as well. The size can range from tens of square kilometers (for example, 45 square kilometers for integrated sustainable water management and

---

Box 2.1   Landscape Definitions: Two Examples

The definitions used by the World Bank and a review article to identify good practices for a landscape approach overlap considerably and cover both the social and ecological dimensions highlighted in the definition proposed by the Landscapes for People, Food and Nature initiative.

World Bank Forest Source Book, 2008
Purpose: to restore forest landscapes

> A landscape is often defined as a geographical construct that includes not only biophysical features of an area but also its cultural and institutional attributes. A landscape is not necessarily defined by its size; rather, it is defined by an interacting mosaic of land cover and land-use types relevant to the processes or services being considered or managed.

Sayer et al. 2013 review article to identify good practices for a landscape approach
Purpose: to reduce conflict (manage trade-offs better) and identify synergies between different land uses such as biodiversity, agriculture, and mining

> A landscape is an area delineated by an actor for a specific set of objectives. It constitutes an arena in which entities, including humans, interact according to rules (physical, biological, and social) that determine their relationships. In many cases, the objectives, arena, entities, and rules will change: our point is that the landscape is defined in broad conceptual terms rather than simply as a physical space.

---

small-scale irrigation schemes to improve household food security in selected areas of Senafae, Eritrea; 95–110 square kilometers for Guassa-Menz Community Conservation Area) to thousands of square kilometers (for example, 2,240 square kilometers for the Weatsbha community in northern Ethiopia reclaiming its land through the reforestation and sustainable soil and water management). Landscape boundaries may be discrete (for example, the ridge defining a watershed) or fuzzy (for example, transition zone between forest and grassland habitats). Landscape boundaries can coincide with administrative boundaries or cut across administrative units (see box 2.2 for examples on how a landscape fits into different ecological scales, compares to other scales, and links to the delivery of ecosystem services and farming practices).

Box 2.2 Ecological and Institutional Scales, Agricultural Interventions, and Ecosystem Services

Figure B2.2.1 compares ecological and institutional scales and how these scales relate to different (agricultural) interventions from field to international level. A landscape is larger in size than a farm and smaller than a region. A landscape can be smaller or larger than a watershed or coincide with a watershed (watersheds themselves can range in size from micro-watersheds to large basins and therefore correspond to the landscape and regional ecological scales shown in the figure). It is important to point out that the ecological and institutional scales shown in the figure do not perfectly align in practice.

Figure B2.2.1 Agricultural Interventions and Ecological and Institutional Scales

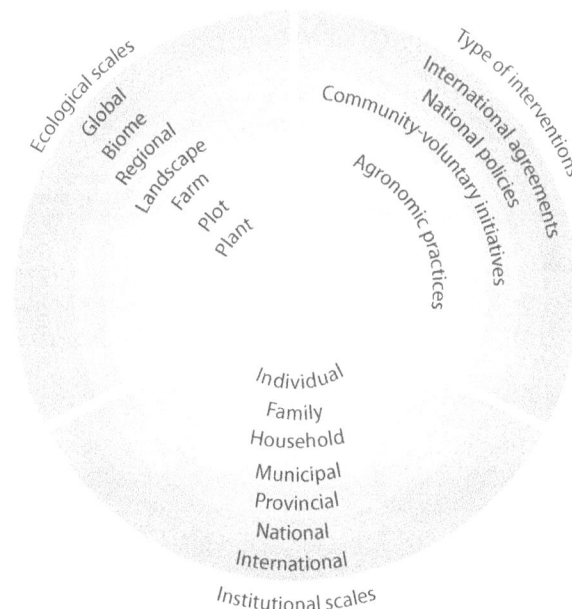

box continues next page

Farmers can promote different farming practices within a field, among their fields, at the perimeter of their fields, or within a landscape (see table B2.2.1). The benefits (including, for example, ecosystem services such as pollination, water-holding capacity, pest control) derived from these farming practices accrue both on farm and off farm in the surrounding landscape. Investments in off-farm practices such as safeguarding forests or meadows, establishing vegetative buffers along waterways, or planting borders along fields can complement and support on-farm practices such as farmer managed natural regeneration and the use of improved planting pits.

Table B2.2.1 Agricultural Interventions and Ecosystem Services at Different Scales

Ecosystem service	Compost or manure	Intercrop	Agroforestry	Insectary strip	No till or low till	Rotation	Cover crop or green manure	Fallow	Border planting	Riparian buffers	Woodlots, meadows, forests
Biodiversity (above and below ground)	♦	♦	♦	♦	♦	♦	♦	♦	♦	♦	♦
Soil quality	♦	♦	♦		♦	♦	♦	♦		♦	♦
Nutrient management	♦					♦	♦		♦	♦	♦
Water-holding capacity	♦		♦		♦		♦	♦			
Weed control		♦	♦			♦	♦				
Disease control	♦	♦	♦			♦	♦				
Pest control		♦	♦	♦		♦	♦	♦	♦	♦	♦
Pollination		♦	♦	♦		♦	♦	♦	♦	♦	♦
Carbon sequestration	♦		♦		♦		♦	♦	♦	♦	♦
Energy-use efficiency	♦	♦	♦	♦		♦	♦		♦		
Resilience to drought	♦		♦		♦		♦	♦		♦	
Resilience to hurricanes/heavy rains	♦	♦	♦		♦	♦	♦	♦	♦	♦	♦
Productivity/yield	♦	♦	♦	♦	♦	♦	♦	♦			
Scale	**Within-field**				**Field**				**Perimeter**	**Landscape**	

Sources: FAO 2011; Kremen and Miles 2012.

Definition of Integrated Landscape Management and Associated Principles of Good Practice

Using a landscape approach is the translation of the landscape concept into analysis, planning, or collaborative processes. Since the purpose and institutional arrangements for managing a landscape can differ greatly (see previous section), the exact definition for a landscape approach will vary (see box 2.3 which highlights different definitions of a landscape approach applied for different purposes). Applying a landscape approach, however, has been referred to as using a concept that is "constructively ambiguous" (people agree on the approach in principle but disagree on specific details that require further negotiations) (Sayer et al. 2013).

Box 2.3 Landscape Approach Definitions: Examples

Relying on the same two sources used to define a landscape and one additional source, a comparison reveals a slight difference in wording for the definition of a landscape approach. For the World Bank Forest Source book, a landscape approach is an analytical tool to optimize land use and interactions among land users within a geographic area. In the Sayer et al. (2013) review article, a landscape approach aims at resolving competing land-use interests. TerrAfrica's definition of a landscape approach (Agostini, personal communication 2014) emphasizes the interconnectedness of different land uses that are managed to increase farm and natural resources productivity and provision of ecosystem services.

World Bank Forest Source Book, 2008
Purpose: to restore forest landscapes

> A landscape approach is a conceptual framework that allows for a structured way of viewing the broader impacts and implications of any major investment or intervention in the rural sector. It describes interventions at spatial scales that attempt to optimize the spatial relations and interactions among a range of land cover types, institutions, and human activities in an area of interest.

Sayer et al. (2013) review article to identify good practices for a landscape approach
Purpose: to reduce conflict (manage trade-offs better) and identify synergies between different land uses such as biodiversity, agriculture, and mining

> Landscape approaches seek to provide tools and concepts for allocating and managing land to achieve social, economic, and environmental objectives in areas where agriculture, mining, and other productive land uses compete with environmental and biodiversity goals.

TerrAfrica, 2014
Purpose: to restore degraded lands and scale up sustainable land management practices

> A 'landscape approach' means taking both a geographical and socio-economic approach to managing the land, water and forest resources that form the foundation—the natural capital—for meeting goals of food security and inclusive green growth. It is done by connecting crop, range, pasture, forest, wood, and protected area lands for provision of ecosystem services and increased productivity.

The Landscapes for People, Food and Nature initiative champions a convergence of these different definitions under "integrated landscape management," a term broad enough to encompass many sector-specific definitions:

> Integrated landscape management refers to long-term collaboration among different groups of land managers and stakeholders to achieve the multiple objectives required from the landscape. These typically include agricultural production, provision of ecosystem services (such as water flow regulation and quality, pollination, climate change mitigation and adaptation, cultural values); protection of biodiver-

sity, landscape beauty, identity and recreation value, as well as local livelihoods, human health and, well-being. Stakeholders seek to solve shared problems or capitalize on new opportunities that reduce trade-offs and strengthen synergies among different landscape objectives. Because landscapes are coupled socio-ecological systems, complexity and change are inherent properties that require management.

This book adopts this definition of integrated landscape management. Integrated landscape management presents an opportunity to scale and leverage land and water management interventions such that the whole is greater than the sum of individual interventions in terms of ecological and economic gains. EcoAgriculture Partners (2013b) identified key actions for operationalizing integrated landscape management to promote successful drylands restoration and community resilience, including: (1) Interventions are designed to promote multiple goals and objectives; (2) Ecological, social, and economic interactions are managed to reduce negative trade-offs and optimize synergies; (3) Roles of local communities are acknowledged; (4) Planning and management of interventions is adaptive; and (5) Collaborative action and comprehensive stakeholder engagement is encouraged and institutionalized. Based on the EcoAgriculture Partners (2013b) report and research results, this book categorizes these key actions into three broad core components for operationalization, and then provides 10 key principles that can be viewed as a checklist for implementation and operationalization of integrated landscape management. As seen in figure 2.1, the three core components are:

Core Component 1: Landscape Goal(s) Encompassing Multiple Objectives at Different Scales

In drylands with mixed land uses and stakeholders, it is important to establish a shared perception of dryland landscapes by identifying and fostering multiple objectives and goals. This promotes common entry points among stakeholders for collaboration for actions that are critical to enhancing resilience. In Sub-Saharan African drylands, goals and objectives generally focus on improving food security and livelihoods diversification. In some higher-producing regions, goals focus on sustainable intensification of production systems. Integrated landscape management must also consider multiple scales at which to implement interventions (for example, farm level and landscape level) as well as temporal and biophysical dimensions—which are especially important in environments with highly variable and unevenly distributed rainfall.

It is important to generate short-term economic returns to incentivize farmer and herder participation, but integrated landscape management also promotes thinking holistically about maximizing ecological gains, for example by improving biophysical connectivity to restore groundwater levels or by providing critical corridors for livestock movement and wildlife habitat. Integrated landscape management must also consider the multi-functionality of landscapes and provide a mechanism to enable local stakeholders to reduce conflicts among different types of specialized resource users (such as herders, farmers, fishers) who differ in their dependencies on a range of ecosystem services.

Figure 2.1   Core Components of Integrated Landscape Management

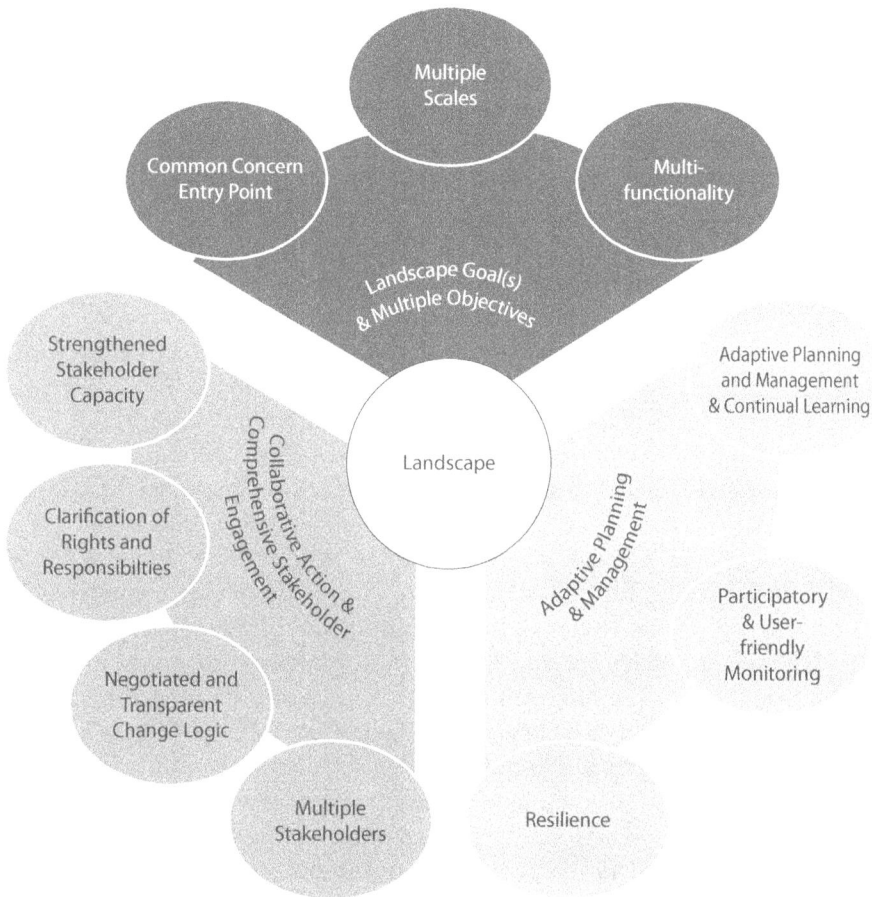

Source: Based on Sayer et al. 2013.

Core Component 2: Adaptive Planning and Management

Integrated landscape management must seek to understand how land users interact with their environment and key sources of income that can improve welfare. The planning of land use, grazing, and natural resource use under integrated landscape management must recognize ecological, social, and economic interactions among different parts of a landscape, which then can be managed to optimize synergies and reduce negative trade-offs. Integrated landscape management should promote continual learning from outcomes and create opportunities to scale successes and address failures. Adaptive management is also important for understanding the resilience of a landscape or how it responds to shocks such as changes in rainfall and temperature. As climatic and economic risks create uncertainty, adaptive planning and management, whereby stakeholders review at recurring intervals the successes and challenges of current land use choices, allows all involved to quickly address risks. As such, integrated landscape management requires effective user-friendly participatory monitoring and evaluation systems and feedback mechanisms.

Core Component 3: Collaborative Action and Comprehensive Stakeholder Involvement

Integrated landscape management must recognize that it is critical to identify and acknowledge local communities and households' roles in resource management. Integrated landscape management must promote community-wide participation in drylands restoration projects and planning, collective action for implementation of restoration interventions, and coordination among key stakeholders across scales and sectors. For example, the collaborative actions of farmers on steep slopes in tandem with concerted actions by herders to reduce grazing pressures in critical locations will have greater impact on erosion and sedimentation rates and restoration of the vegetative cover than fragmented or individual efforts alone. Local communities must be incentivized to invest in improved land and water management and share local knowledge and experience.

To increase the likelihood of long-term success, implementation of integrated landscape management must build on lessons learned from past experiences. Sayer et al. (2013) identified 10 principles of good practice based on a comprehensive review of published literature on landscape approaches and a consensus-building process. These principles overlap with good practices identified for successful collective action to provide environmental public goods and reduce negative externalities in agricultural landscapes (OECD 2013). These principles are also in line with minimum standards and good practices for supporting sustainable pastoral livelihoods developed by IUCN, the World Initiative for Sustainable Pastoralism, and others (IUCN-ESARO 2012). The 10 principles are useful to design a process that motivates various stakeholders to pursue a common goal within a landscape, make synergies and trade-offs between objectives more transparent, and establish agreed upon mechanisms to resolve differences between stakeholders.

The following chapters on the role of stakeholders, the ecological and economic assessment of cost and benefits, and the review of case studies all look systematically at these 10 principles of good practice. Figure 2.1 groups these 10 principles into three broad categories, and table 2.2 highlights examples and explains the 10 principles more in depth.

Table 2.2 Principles of Good Practice for Integrated Landscape Management: Examples

Principles of good practice	Explanation and examples
Landscape goal(s) and multiple objectives	
Common concern entry point	• Establish a shared perception of landscape as a management unit. • Set clearly defined, achievable, short-term goals related to a landscape and work on issues incrementally to build trust (rather than a "perfect" complex landscape goal). • Seek reasonable and achievable outcomes (for example, restore watershed, safeguard road from floods).

table continues next page

Table 2.2 Principles of Good Practice for Integrated Landscape Management: Examples *(continued)*

Principles of good practice	Explanation and examples
Multiple scales	• Achieve outcomes (at least) at two scales: farm level (household) and landscape level. • Consider that different problems and interventions will require working at multiple scales nested within each other (for example, micro-watersheds to restore and invest in sustainable land and water management infrastructure and organize effective collective action; a much larger basin to safeguard a road from flooding or ensure it is not washed away with a different set of stakeholders). • Integrated landscape management is generally applied at a scale between a targeted watershed and a large river basin (a geographic area ranging from tens to thousands of square kilometers).
Multi-functionality	• Reflect multiple uses and purposes (valued differently by different stakeholders) such as food production, timber and fiber production, urban habitat, wildlife and biodiversity, and recreation. • Be cross-sectoral, multidisciplinary, and integrated, based on assessing multiple types of land use, production systems, and socio-economic activities in a designated geographic area. • Take account of and help to assess and manage trade-offs (to optimize multiple services instead of always seeking to maximize a given production function or service) among a range of provisioning, regulating, and supporting ecosystem services.

Adaptive planning and management

Adaptive planning and management and continual learning	• A fundamental element of integrated landscape management is learning from outcomes and creating opportunities to adjust a landscape strategy and associated actions because many ecological production functions are non-linear; natural and social systems may shift suddenly with thresholds; external shocks are difficult to predict; and cause and effect of different landscape processes may be poorly understood.
Participatory and user-friendly monitoring	• Ensure that information on the progress of landscape actions and associated benefits (at all relevant scales) is readily available. Participatory systems need to be in place to generate such information, and stakeholders must be able to integrate and analyze this information.
Resilience	• Maintain landscape attributes (and associated social systems) that help to avoid or deflect shocks and support households to absorb and recover from shocks (for example, scale sustainable land management practices, better adapted and more resilient production and land-use systems).

Collaborative action and comprehensive stakeholder engagement

Multiple stakeholders	• Involve a range of management objectives, including both production-oriented and conservation and ecosystem management-based objectives, established by multiple stakeholders that seek multiple benefits across space, time, and sectors. • Design a process to achieve multiple "wins" including increased productivity, greater food and water security, enhanced adaptation and resilience, greenhouse gas mitigation, biodiversity conservation, conflict reduction, and other benefits.
Negotiated and transparent change logic	• Promote a transparent process and understanding of what actions will be taken once milestones have been achieved (for example, a community agrees first to restore a watershed requiring specific actions by each stakeholder; once the groundwater table has been restored, the groundwater needs to be allocated in a fair way and its use monitored).
Clarification of rights and responsibilities	• Define stakeholder contributions to different landscape actions, agree upon rules to share benefits, decide on sanctions to avoid free riders, and establish fair mechanisms for conflict resolution.
Strengthened stakeholder capacity	• Provide a platform for dialogue, sharing experiences, and learning, so that all can build capacity. Implementing an integrated landscape management initiative requires dedicated efforts to strengthen social, financial, and other capabilities and build new stakeholder capacities to participate.

Sources: Sayer et al. 2013; Milder et al. 2014; EcoAgriculture Partners 2013b.

What Is Different about Integrated Landscape Management?

Skeptics of a landscape approach to manage natural resources may characterize this quest for more integration across multiple sectors and stakeholders and greater emphasis on geographic targeting as nothing new—this has been tried before under different names (see table 2.1). However, the way a landscape approach is being proposed conceptually here is different (see section on definition and landscape principles). Integrated landscape management builds on lessons learned from previous approaches (table 2.1) and places a larger emphasis on building resilience to drivers like climate change and changing market forces. Integrated landscape management emphasizes principles based on lessons learned from these previous approaches, but also provides value added in that it:

- Does not promote a "one-size-fits-all" approach but rather asks stakeholders to consider the local context and include sectors, stakeholders, and social, cultural, and other conditions into account across geographic boundaries that make ecological sense.
- Emphasizes that planning and implementation take into account spatial components important to rejuvenating and maintaining ecosystem health (for example, hydrological flows, habitat). Integrated landscape management requires that land-use planning and decision making think differently about scale and take into account these spatial components.
- Promotes a combination of bottom-up and top-down principles designed to encourage local community participation, but at the same time dedicated to building appropriate institutional and financial support.
- Promotes an adaptive management approach that tries to build on monitoring and evaluation to generate long-term data needed to truly understand whether communities are becoming more resilient and increasing their adaptive capacity, and whether landscape-level changes are achieved.

Benefits and Risks Associated with Integrated Landscape Management

Integrated landscape management strives to align household practices of land, water, and other natural resource use and spatial arrangements of different land cover types in a way that boosts synergies and reduces negative trade-offs among the supply of goods and services provided within that landscape while buffering households from the effects of climate change. This vision of managing a landscape in a holistic fashion incorporating all land uses within that geographic area and embarking on a systematic long-term effort to seek solutions linked to the landscape scale is the new element compared to past approaches such as the sustainable land and water management approach, which historically focused on land and water management practices using farmers and other resource users as entry points (see table 2.1). Integrated landscape management can only add value if it overcomes observed shortcomings of past sectoral approaches (see next section) and improves upon past integrated and geographic approaches (see table 2.1). Both government and international development cooperation actors support such a vision of integrated landscape management as the following examples indicate:

- "Functions within a landscape are all interconnected socially as well as bio-physically. But hitherto, they've been managed in isolation. What we really need is a much more holistic approach to understand these interconnections, to capture their complexity, and to make sure their management is integrated and simplified in a way that hasn't been done before." The Ministry of Economic Affairs, Netherlands.
- "Meeting international goals for food security and inclusive green growth requires better integrating the management of land, forests, and water resources. Doing so will help to maximize productivity, improve livelihoods, and reduce negative impacts on the environment." The World Bank.

This section summarizes the following major benefits of a landscape approach that can be delineated from a conceptual perspective: geographical and ecological-scale benefits, cost savings resulting from economies of scale or scope, and capacity development and learning from collective action (OECD 2013). It concludes with examples highlighting the risks associated with integrated landscape management.

Geographical and Ecological Scale Benefits
A landscape approach will be advantageous if long-term solutions to a problem require understanding of the spatial processes within a geographic area, if a landscape is the right scale to organize collective action to provide public goods or to reduce externalities, or if a landscape-level effort results in overcoming an ecological, biophysical, or social threshold to provide a minimum supply of an ecosystem service or other benefit. For example, these benefits are likely to be generated when:

- *Ecological, biophysical, and socio-economic characteristics and interactions within a geographic area require a spatial approach.* Irrigation water, soil erosion, wildlife, and pastoral livestock movements, for example, all have to be managed at the appropriate geographical and ecological scale to be sustainable over the long term. Investments in soil and water management or landscape restoration will not produce maximum benefits if they do not reflect agro-ecological gradients, hydrogeology, soil conditions, and socio-economic conditions of an area (van Steenbergen, Tuinhof, and Knoop. 2011). Programs and investments have to be based on correct geographical boundaries of natural resources such as watershed or aquifer extent. If a degraded watershed, for example, cuts across multiple administrative units, it is necessary for all administrative units to be stakeholders in the watershed restoration effort. Similarly, a landscape may be the right socio-economic scale—"be local enough and large enough in size"—to solve a local problem that requires specific local knowledge and expertise, which could not be addressed by more general national policymaking or by market mechanisms (OECD 2013). In Tanzania, for example, restoring woodlands and dry season grazing areas for livestock through assisted natural tree regeneration required collaboration among communities covering a large geographic area and has helped households to diversify livelihood strategies and helped buffer dry season risks for livestock (WRI et al. 2005).

- *Long-term solutions to a problem can only be implemented successfully at a landscape scale.* In this case, collective action at a landscape scale is needed to provide public goods or reduce externalities. Examples of such public goods in agricultural landscapes include restoring wetlands and riparian areas for their pollutant removal services and other benefits. Organisation for Economic Co-operation and Development (OECD) examples for creating such public goods in agricultural landscapes include wetland restoration in the Mulgrave River basin discharging into the Great Barrier Reef lagoon in Australia or long-term land management contracts in the Eider Valley in Germany that support extensifying land use (establishing collective wetland grazing areas), removing existing drainage systems, and establishing new flood zones (OECD 2013). In Kenya, providing incentives to leave wildlife migration routes in the Kitengela Plains unfenced has created new income for dryland farmers and improved wildlife, biodiversity, and tourism benefits for Nairobi National Park (Kristjanson et al. 2002; Gichohi 2003).
- *Similar coordinated action may be required to reduce negative environmental externalities, such as the reduction of non-point nutrient pollution from farms or conflicts between pastoralists and crop farmers.* In three regions in Niger, for example, demarcated livestock corridors have protected farmers' crops and trees, safeguarded grazing and water access areas for herders, and resolved conflicts through agreed upon dispute resolution mechanisms (Byrne et al. 2011; Learning Initiative 2012). Although Niger's Rural Code (1993) and revised Pastoral Code (2010) protect livestock corridors, funding and technical support at landscape scale overcame lack of government resources and facilitated collaboration among farmers, pastoralists, local and regional governments, and NGOs to reduce the incidence and intensity of conflicts and assist livestock movements to markets.
- *A landscape scale may be required to link the right set of actors and beneficiaries to ensure long-term success of changes in farming practices and financial sustainability.* OECD examples for reducing negative externalities in agricultural landscapes include the Pyhäjärvi Restoration Project that seeks to prevent eutrophication of Lake Pyhäjärvi in southwest Finland or the effort to recycle drained water from irrigation agriculture in the Shiga Prefecture to reduce the flow of agrochemicals into Lake Biwa in Japan (OECD 2013).
- *A landscape approach is needed to provide a minimum supply for an ecosystem service or benefit (threshold effect).* Some ecological production functions and public goods in a landscape exhibit non-linear changes when they reach

certain thresholds. A small variation in an independent variable can produce a large abrupt change in a dependent variable such as the supply of an eco-system service or public good. Ecological systems function differently after they reach such a threshold, affecting their resilience (this is referred to as a regime shift, for example small changes in cattle densities in wet savannas have shifted grass-dominated rangelands to more woody savannas) (Ander-ies, Janssen, and Walker 2002). These public goods are sometimes referred to as "threshold public goods" or "non-linear public goods" (OECD 2013). Col-lective action in a landscape can be managed to actively cross such thresh-olds to achieve a large boost in an ecosystem service or to avoid a large loss of an ecosystem service, which would be detrimental to the stability of a farming system, livelihoods, or economic development (for example, perma-nent loss of all top soil).

- *Efforts to control a livestock disease or manage groundwater resources exhibit such threshold effects.* Some livestock diseases, for example, can be controlled within a geographic area, after a sufficient number of farmers have consis-tently applied prophylactic measures or an area-wide removal of a vector has been conducted (for example, tsetse vector). Aquifers can collapse when too much water is removed or become unusable once saltwater or other con-taminants have entered. Similarly, watershed restoration in Tigray, Ethiopia has shown that once upper catchments were rehabilitated (following a catchment logic from ridge to valley), associated groundwater and economic benefits were achieved in the lower catchment. Once groundwater tables were restored to a certain level, farmers could access the water with treadle pumps or hand-dug wells. This allowed small-scale farmers to shift from unreliable rain-fed agriculture to dry season irrigation, with a concomitant shift to higher value crops such as vegetables and fruit trees (Hagazi and Hailemariam 2012).

Other examples showing threshold gains or losses in benefits include:

- Households in southern Niger report (Larwanou, Abdoulaye, and Reij 2006) reduced wind speed at the beginning of the growing season after they in-creased on-farm tree densities (a certain minimum tree density and spatial distribution are required to achieve noticeable effects).
- Invasive species such as *prosopis juliflora* in commercial farms along the Awash River in northeastern Ethiopia (Behnke and Kerven 2011; Learning Initiative 2012) block pastoralists' access to the Awash River during the dry season or drought and could reach a level where access is no longer possible.
- Landscape design for biological control (Tittonel 2013) requires fine-tuned ecological engineering to provide alternative sources of food and shelter for natural enemies distributed strategically in space and time. In Ethiopia, for ex-ample, the International Maize and Wheat Improvement Center (CIMMYT) is testing new approaches to control maize stem borers.

Other public goods within a landscape with threshold effects include: level of biodiversity and the supply of ecosystem services; amount of nutrient runoff or nutrient concentrations in freshwater bodies and frequency of algal biomass (and algal blooms); level of impervious surface in a watershed and downstream water quality; and grass density and the frequency of fire. Land-care associations (coalition of farmers and other land users) in OECD countries, for example, have coordinated their actions to organize a sufficient number of land users to provide non-linear public goods such as diverse agricultural landscapes with high aesthetic and biodiversity benefits (OECD 2013).

Cost Savings Resulting from Economies of Scale or Scope
By sharing their skills and assets, farmers and other land and water users in a landscape can achieve economies of scale—that is, cost advantages resulting from bringing stakeholders together across broader geographic scales and sectors. For interventions with high fixed costs such as building large-scale irrigation infrastructure and safeguarding water supplies, cost savings could be considerable (OECD 2013). Some landscape interventions can also provide economies of scope—cost advantages resulting from generating two or more products simultaneously (OECD 2013). The effort to coordinate different land users in an area (for example, farmers close to rivers, farmers close to wildlife habitat) can be designed to optimize the supply of multiple ecosystem services (Dosskey et al. 2012) within a landscape (for example, improving water quality, increasing wildlife populations, and sequestering carbon). In addition to cost savings, integrated landscape management can generate a multitude of co-benefits. Direct and co-benefits are explored more in depth in chapter 4.

Capacity Development and Learning from Collective Action
An OECD review of collective action in agricultural landscapes to provide public goods demonstrates that collective action can enhance farmers' expertise. This in turn can harmonize objectives among the actors within a landscape, attract additional support, and build new capacity to cope with future change (OECD 2013). The capacity development benefit of collective action in a landscape becomes more pronounced, once learning and feedback among participants shifts from an informal and *ad hoc* approach to a more deliberate integrated and continuous process.

Risks Associated with Integrated Landscape Management
Integrated landscape management can be an effective strategy for drylands. It should not be seen as a "one-size-fits-all" solution but rather a flexible framework for scaling investments at a landscape scale to maximize ecological, economic, and social synergies and minimize negative trade-offs.

Since there is still limited implementation experience with the new aspects of integrated landscape management, especially how to translate all the principles of good practice, there is a risk that new integrated landscape management programs create their own set of shortcomings such as technocratic-driven processes by outsiders and unbalanced approaches lacking local ownership. Much can be learned from the mistakes of past spatial and integrated approaches such as with the *Groupement Européen de Restauration de Sols* (GERES) in the Central Plateau of the Yatenga Region in Burkina Faso (Marchal 1979) in the early 1960s, which invested heavily in mechanized approaches to install physical barriers to water erosion in an effort to restore the productivity of cropland, but which failed to engage local communities in understanding the rationale for these investments or in contributing to the maintenance of these investments.

To mitigate this risk, new integrated landscape management programs need to pay special attention to taking stock of local innovators, empowering and enabling local actors, strengthening community-based institutions, and reinforcing good governance at all levels (accountability, transparency, as well as participatory, equitable benefit sharing). In addition, since a landscape approach often seeks to optimize both public and private goods, experimentation and innovation are needed in regards to new arrangements for private-public partnerships.

It is important to point out that integrated landscape management has additional costs (such as transaction costs for collective action)—which in many locations may be offset by the benefits, especially with a more long-term perspective (see more detail in chapter 3). In some situations, investing in integrated landscape management may not be the favored strategy for selected stakeholders (Kissinger, Brasser, and Gross 2013). A beverage distributor, for example, has a strong interest that the water and the watershed in the sourcing area of bottling plants are managed sustainably. If the company does not own the bottling plant and can easily switch suppliers, it may prefer to rely on bottling plants within multiple watersheds and not make a long-term commitment to a specific watershed (which may require taking on costly watershed restoration and institution building efforts). Kissinger, Brasser, and Gross (2013) discuss in more detail how companies can assess supply chain risks and determine whether integrated landscape management is a viable strategy to tackle a specific risk.

Sectoral versus Integrated Landscape Management Interventions

The majority of government agencies and their plans and budgets are still organized by sectors such as agriculture (sometimes split into livestock and crop subsectors), water, forest, wildlife, and health. Even environment or natural resources management interventions often remain in their own silos and are not always well integrated with the more dominant sectors of the economy. Compared to a more integrated landscape approach, narrow sectoral approaches can be associated with shortcomings as the following examples demonstrate:

• *Too narrow focus in agriculture sector.* Many agricultural interventions emphasize commodity-based approaches (for example, selected commodity value chains) with a strong focus on field-level productivity favoring quick fixes (for example, new seeds, high agrochemical inputs, irrigation). They do not fully account for a farm's dependence and impact on the surrounding agricultural landscape (and associated ecological and social processes). A value chain analysis, for example, could identify favorable gross margins and rates of return for producing green beans with the help of small-scale irrigation. Scaling green bean production from a few to many farmers within a large area may ultimately fail in the long run if water supplies for irrigation are not safeguarded. Similarly, investments in long-term restoration and maintenance of soil resources, both on and off farm, require a broader and more long-term perspective, which are often not part of the standard sectoral technology package—but may be essential for the efficacy of fertilizer applications as observed in Malawi (Winterbottom et al. 2013).

- *Too narrow focus in water sector.* Water managers may still overemphasize abstraction of surface and ground water (blue water) and not consider sufficiently managing water in soil and vegetation (green water), which can represent about two-thirds of total rainfall in many areas (Falkenmark and Rockström 2006). Water sector specialists may propose a catchment restoration strategy with hydrological improvements, but fail to fully assess and communicate the cost and benefits of implementing this strategy to farmers (which is another sector).

- *Too narrow focus in ecological or hydrological restoration efforts.* Sectoral blinders can also be found in ecological or watershed restoration efforts that are too fixated on ecological or hydro-geological processes. Such interventions are associated with a high risk of failure if they do not reflect adequately social processes across land uses or account for economic and social synergies between farms and between different types of land uses (The Planning Commission of the Government of India (2012), for example, mentioned "flawed solutions" and "ridge-to-valley fundamentalism" in its final report of minor irrigation and watershed management for India's twelfth five-year plan).

- *Many sector interventions are implemented without adequate community participation.* Large-scale watershed restoration using top-down approaches without meaningful community participation and rigid technical packages have resulted in failures in numerous countries (for example, see Lakew et al. (2005) for watershed restoration in the 1980s in Ethiopia). Community participation has been shown to increase the effectiveness and longevity of landscape interventions.

- *Many sector interventions disregard fundamentals of nature.* Water infrastructure investments without rehabilitation of upstream watersheds or water harvesting programs disregarding community participation and an integrated approach following a watershed logic are typically mentioned as failing examples of such a narrow approach. The Ethiopia Community-based Participatory Watershed Development Guideline (Lakew et al. 2005) cites the case of the Borkena Dam in South Wello, which was completed in the 1980s before sufficient watershed conservation measures were established. Within one rainy season, the multi-million Birr dam was filled with silt and coarse materials. Similarly, a case study of the Managing Environmental Resources to Enable Transition (MERET) program in Tigray contrasted the successful integrated watershed restoration efforts under MERET with efforts to establish micro-dams covering up to 100 hectares and farm-level ponds to collect water in the same area, which were unsuccessful because they "did not build on a community-oriented and integrated approach" (Hunger, Nutrition, Climate Justice 2013).

Sectoral approaches are attractive because they are well-defined and more predictable in the short term. Table 2.3 highlights some of the main differences between a sectoral and a landscape approach. A Landscapes for People, Food and Nature initiative review observed that sectoral actors are expected to be major policymakers and investors affecting rural landscapes in the near future and therefore proposed a strategy to sufficiently broaden sectoral investments to accommodate a landscape perspective (Kissinger, Brasser, and Gross 2013). This is especially important since a review of integrated landscape approaches in Latin America and Africa found that a landscape approach with its more deliberate stakeholder engagement process is more likely to discover hidden co-dependencies between social and natural systems, less likely to exclude important actors, and more likely to achieve a larger number of positive outcomes, assuming the initiatives aim for multiple objectives (Kissinger, Brasser, and Gross. 2013; Milder et al. 2014; Estrada-Carmona et al. 2014).

The same review also observed that a landscape approach differs from an intervention targeting a large number of farmers across an extensive geographic area. For example, with the help of new value chains linking farmers to markets, the latter intervention could result in considerable landscape benefits. Thousands of farmers applying drip irrigation or integrated pest management, for example, would save water for other water users or positively affect pollinators within the landscape. Using markets, farmer field schools, or extension services to scale farming practices can achieve positive landscape-level benefits, especially when the promoted farming practices involve more efficient use of land, water, or agricultural inputs. The collective action and the landscape-level benefits, however,

Table 2.3 Differences Between a Sectoral and a Landscape Approach

	Sectoral	*Landscape*
Scale	Local: 1 or 2 land uses	Larger scale: multiple interacting land uses; fuzzy or discrete
Timescale	Short to medium term (1–5 years)	Many years to several decades
Scope	Well-defined	Well-defined but adaptable
Management	Clear and well-defined organizational roles and structures	Roles evolve and overlap; civil society has increasing significance
Learning	Project-cycle level; can be informal	Integral and continuous
Authority	Centralized and clear	Decentralized and distributed; negotiated

Source: Sayer et al. 2013.

are not coordinated. The authors suggested that "without intentional coordina-tion, landscape benefits may emerge only coincidentally but are far from certain, and opportunities for additional synergistic cost-savings and concurrent impacts may be missed altogether." This may be especially important when stakeholders are trying to address multiple complex risks, such as poverty and climate change adaptation (Kissinger, Brasser, and Gross 2013).

Relevance of Integrated Landscape Management to Drylands

Dryland communities along with their production systems and human livelihood strategies have evolved over hundreds of years in response to an unfavorable climate, enabling both ecosystems and human well-being to recover following droughts, floods, and fires. Over the past decades, however, high human popula-tion growth rates, land-use pressures and land degradation, greater frequencies and intensities of droughts, conflicts, and other natural and anthropic drivers have begun to undermine the resilience of many dryland communities in Africa. Local communities are facing a reduced capacity of the land to support them, lowering their resilience to recover from natural shocks. The following challenges are espe-cially relevant from a landscape perspective:

- Depleted soil fertility, water stress, and land degradation.
- Greater exposure to shocks, such as more irregular and extreme rainfall in the Sahel.
- Increased conflicts over land, water, and other natural resources.

Although there are an increasing number of positive experiences, efforts to address these challenges in the drylands of Sub-Saharan Africa have too often failed to achieve significant and lasting improvements at scale. Few interventions have been designed to take into account the linkages between upstream farmers and downstream water users. In many cases, interventions have disrupted tradi-tional management systems for common pool resources such as wetlands, grazing reserves, and forests.

Single-objective and sectoral development approaches in particular are increasingly seen as inadequate because they may not fully address trade-offs associated with competing land uses and actors, or fall short in incorporating the perspectives of all stakeholders in local communities and in appropriately addressing sources of resource conflict. They may also fail to take into account the biophysical connections and leverage interactions among production systems which are critically important in dryland systems and necessary to generate and sustain both farm-level and landscape-level benefits. For example, trees in agri-cultural landscapes play a critical role in renewing soil fertility, providing addi-tional sources of fodder for livestock and fuelwood for households, and sustaining cropland productivity, while simultaneously contributing to the diversification and enhanced resilience of farming systems; yet many agricultural and livestock development programs have not taken full account of the key roles of trees in agricultural landscapes. Many development actors across Sub-Saharan Africa are starting to adjust drylands development programs in such a way that they consider multiple objectives and multiple actors across two or more sectors; applying a landscape approach can increase the effectiveness of these programs and capitalize on opportunities to restore resilience in drylands.

Integrated landscape management is particularly important for drylands as people depend on production systems that are frequently disrupted by exogenous shocks such as drought. Households rely on movement over relatively large areas or on diversification of livelihoods. The latter includes the use of multiple plant and animal species, but also labor migration and transformation of and marketing of agricultural products. This reliance on multiple ecosystems and ecosystem services can become easily unbalanced with an isolated management focus on one sector or one commodity, which in turn may degrade other resources, reduce food security, or increase other risks. Landscape approaches can help to avoid or minimize the potentially negative impacts of such uncoordinated, sector-specific interventions and capitalize on potential synergies.

Resilience and Integrated Landscape Management

This section introduces the three dimensions of household resilience and explores how different dimensions of resilience are linked to actions at household and landscape levels. It concludes with dryland-specific opportunities to reduce vulnerability and increase resilience with the help of integrated landscape management.

Household and Landscape Interventions to Increase Resilience in Drylands

This book is part of a set of more comprehensive World Bank books that aim to support dryland countries in Africa to achieve food security, create employment and income opportunities, strengthen their social fabric, and build their natural capital to ensure sustainable growth (World Bank 2013). Drylands, defined by their length of growing period, aridity level (lack of water/moisture), and variability of rainfall, cover about 43 percent of Africa's land area (including arid, semi-arid, and sub-humid zones), and encompass about 75 percent of Africa's agricultural land used for cropping and livestock (World Bank 2013). In 2010, about half of the Sub-Saharan African population lived in drylands, which are expected to be the home of an additional 350 million people by 2030 (World Bank 2013). Drylands generally are among the poorest and most marginalized areas of Africa and can be risky environments for conventional investments aimed at increasing the production of rain-fed crops. Most dryland countries in Africa are highly vulnerable to external shocks posed by climate (for example, drought, floods), diseases affecting humans, animals and plants, price changes buffeting food or export commodities, and social conflict.

The set of more comprehensive World Bank books seeks to describe and quantify the relationship between productive capacity, external shocks, and development outcomes in dryland areas (World Bank 2013). For that purpose, the World Bank established an analytical framework to assess vulnerability and resilience at different scales (national, sector, community, and individual) covering three different dimensions:

- *Exposure to shocks*: A household relying on growing rain-fed maize for its livelihood in a semi-arid area has high exposure to drought, for example. Similarly, economies relying on rain-fed agriculture in dryland areas such as Ethiopia and Kenya witness high exposure to drought and flood risks, which then is reflected in their annual fluctuation of national gross domestic product (GDP) (Mogaka et al. 2005; World Bank 2006b; Conway, Lisa, and Schipper 2011).
- *Sensitivity to shocks*: A household with a higher proportion of its livelihood or a country with a greater share of its economy dependent on shock-affected activities such as pasture-based livestock production or rain-fed cropping is expected to be more vulnerable to drought-related shocks.
- *Coping capacity*: A household with greater financial capital or with a comprehensive social network providing emergency support can cope better with shock. Similarly, countries with well-functioning early warning systems, a good road network, sufficient emergency funds, and a responsive social safety system will be better able to buffer the impacts of a shock and support affected households to recover more quickly.

The total vulnerability of a household to shocks (and its inverse, the resilience to bounce back from shocks) is the product of the household's level of exposure to shocks, the sensitivity by which its assets and income are affected by the shocks, and its coping capacity to mitigate the effects of the shocks. Households can reduce their vulnerability to shocks by increasing their coping capacity that mitigates the impacts of the shocks on their livelihoods, income stream, or assets, for example by accessing savings or loans. Farmers can reduce their sensitivity to shocks by shifting their livelihood's dependence on activities or assets that are less sensitive to the shock (for example, drought-tolerant crop varieties or livestock species more resilient to disease vectors). A rural household can reduce its exposure to drought-related shocks, for example, by creating off-farm and non-farm opportunities for its livelihood or deciding to migrate to urban areas.

Increasing coping capacity that requires no major changes in livelihoods, but mitigates the effects on household assets or income is easier to implement than efforts to reduce exposure that create new livelihoods. Over the long term, reducing the exposure to a specific shock is more effective. Figure 2.2 provides examples of different options to reduce vulnerability with household-level interventions and summarizes how they could be prioritized across dimensions of resilience depending on their level of long-term effectiveness and their ease of implementation.

Figure 2.2   Household-Level Interventions and Dimensions of Resilience

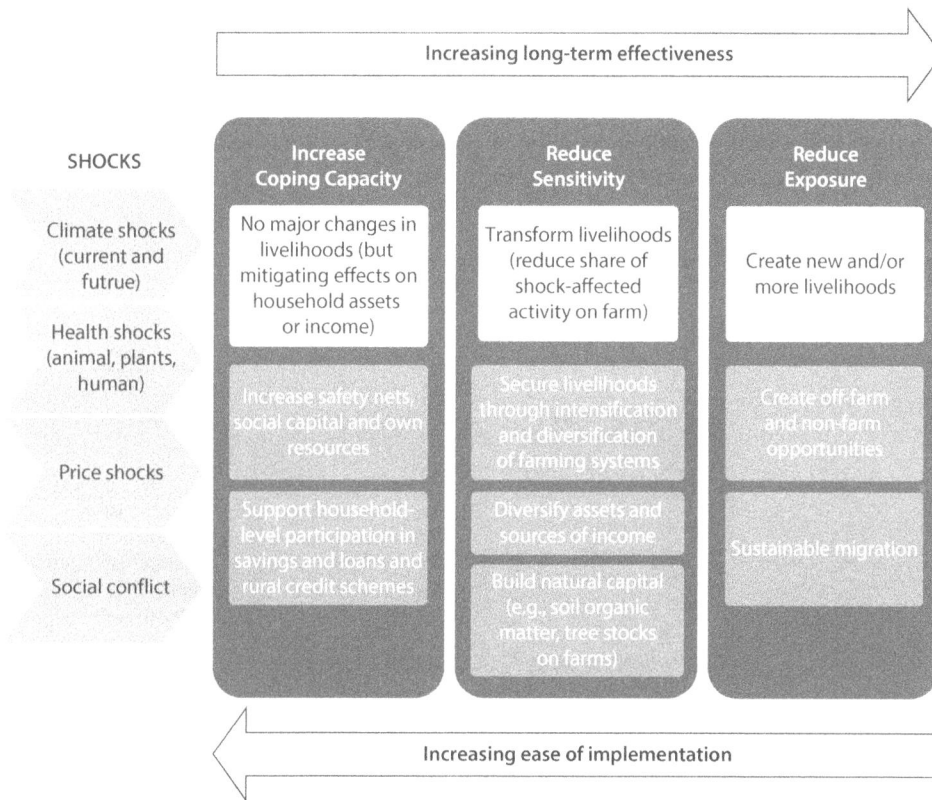

Source: Adapted from a presentation slide by R. Cervigni, World Bank 2013.

Rural households working through community-based organizations engaged in integrated landscape management can prepare for and respond to shocks by changing the activities on their farms and other land uses or by building their social capital. They can also manage their risk by taking advantage of the matrix of different land uses and availability of common pool resources within a landscape. Increased investments in protecting and regenerating trees and shrubs on farms and in restoring pastures and forests and other common pool resources within a landscape, for example, can reduce the risk for households resulting from external shocks (Dewees 2013). Studies examining the use of trees on farms and the reliance of rural households on forests and other common pool resources have shown that households use these additional tree and forest resources including wood, fodder, fruits, edible leaves, and other products to diversify their livelihood strategies and reduce their risks.

Vulnerability to shocks can also be reduced by landscape-level efforts involving collaborative or collective action by a group of farmers and other land and water users. These interventions can range from efforts that seek no major changes in the distribution of land-use and land-cover types within a landscape, to transformation of existing land use and land cover (the same distribution of land-use and land-cover types within a landscape, but a change in management intensity for each type), and finally to fundamental structural shifts in the distribution of land use and land cover within a landscape.

Following the same conceptual approach as in figure 2.2, figure 2.3 shows examples of landscape-level interventions that reflect a gradient in ease of implementation and increase in long-term effectiveness.

Figure 2.3 Landscape-Level Interventions and Dimensions of Resilience

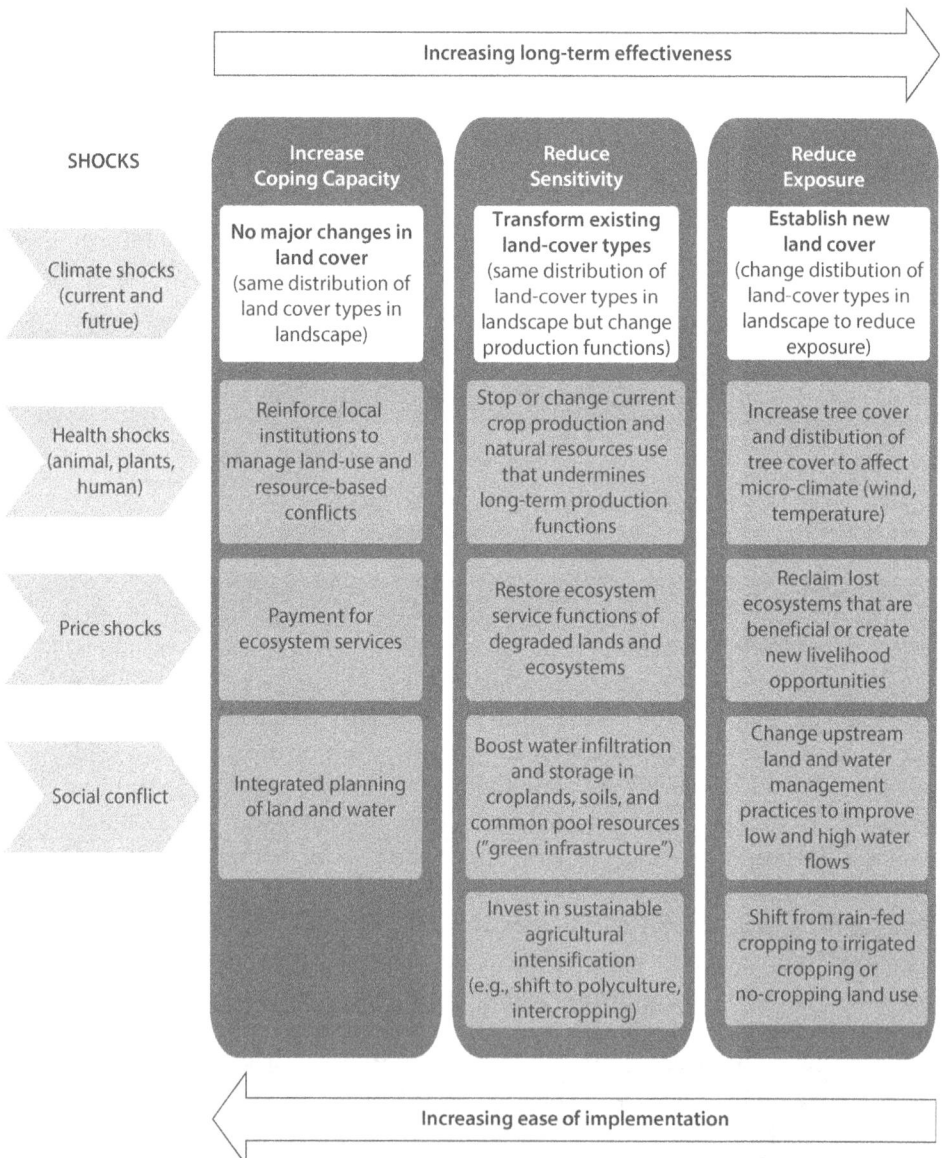

Source: Adapted from a presentation slide by R. Cervigni, World Bank 2013.

Figure 2.3 shows examples of policy interventions that could increase cop-ing capacity within a landscape: strengthening community-based institutions governing access and use of natural resources, providing land users with rewards for providing public non-market goods (ecosystem services), improv-ing more integrated planning and coordination, or establishing more effective mechanisms to resolve land and resource-based conflicts. Policies can also support collaborative or collective action within a landscape that reduces sensitivity to shocks by investing in agriculture that protects and stabilizes existing ecosystem functions (for example, maintain hydrological base flows or increase groundwater recharge), with the help of agricultural practices that increase water-holding capacity in soils or change surface runoff coeffi-cients. Most effective changes within a landscape over the long term, but more difficult to implement, are structural shifts in land cover and land use, which could include reclaiming degraded lands, scaling up sustainable agri-cultural intensification, and phasing out detrimental land uses. For example, commercial plantation forestry is classified in South Africa's Water Act as a stream flow reduction activity. This has resulted in land-cover changes in water-constrained watersheds with tree-free buffer zones being established close to permanent streams and public works programs removing invasive tree seedlings to improve stream flow (Cadman et al. 2010; Albaugh, Dye, and King 2013). Similarly, dismantling inappropriate enclosures and reopen-ing former stock routes may be required to ensure pastoral mobility and increase resilience of pastoral production systems (Wagkari 2009; IUCN 2012).

Dryland-Specific Opportunities for Integrated Landscape Management to Increase Resilience

Water scarcity, land degradation, and loss of biodiversity are major biophysical constraints facing drylands and are key threats to economic development and human welfare. Sustainable land and water management interventions to con-serve soil and water, build natural and social capital, and maximize efficiency of water and soil resource use serve as a good foundation for integrated landscape management in that they help to rebuild household resilience and are critical for stabilizing and intensifying rural production systems.

The following practices have been identified as especially promising (Winterbottom et al. 2013) for drylands where the need for the widespread adop-tion of improved land and water management practices to boost productivity is especially acute: agroforestry, farmer-led soil and water conservation techniques, rainwater harvesting, conservation agriculture, and integrated soil fertility man-agement. These measures have been effective in reversing land degradation and in contributing to the sustainable intensification of agriculture and forestry. Rural economies benefit from these practices through higher crop yields, increased sup-plies of fodder, firewood, and other valuable goods, greater income and employ-ment opportunities, a restoration of biodiversity and ecosystem services, and higher climate change resilience. These practices can be scaled up to enhance and diversify production systems and increase household resilience.

Sustainable land and water management has historically focused on aligning land capabilities with improved land and water management using farmers' and other resource users' practices as entry points (see table 2.1). To increase resilience for the greatest number of households in drylands, however, integrated landscape management must take sustainable land and water management further by:

- Identifying non-sustainable land-use practices that contribute to the degradation of the natural resources that are the foundation of food production systems and rural livelihoods, looking beyond traditional project scales to incorporate actions across a wider geographic context.
- Encouraging participation by local user groups in the planning process and encouraging broader participation by bringing new stakeholders to the table.
- Facilitating dialogue with local user groups to consider viable options to scale up such improved management practices over a wider geographic area.
- Considering the dominant land uses (measured by the number of people or area) such as rain-fed cropping and livestock production in each landscape.
- Seeking to solve shared problems or capitalize on new opportunities that reduce trade-offs and strengthen synergies among different land-use options within a landscape.

Organizations such as the Food and Agriculture Organization of the United Nations, TerrAfrica, the African Union, and NEPAD are incorporating more integrated landscape management principles into their definitions of sustainable land and water management, such as promoting multi-level and multi-stakeholder involvement, participatory approaches, and integrated use of natural resources at ecosystem and farming systems levels (FAO 2014).

Opportunities to restore resilience in drylands will vary by country and among landscapes. The specific entry points and goals for integrated landscape management will vary. In many situations sequencing of activities may be the best approach. This implies starting as simple as possible and allowing for growing complexity based on the evolving needs and priorities of land users. The following represent dryland-specific opportunities for integrated landscape management:

- *Restoring degraded drylands can become the entry point for extending integrated landscape management.* Integrated landscape management is well suited to address the major biophysical constraints to the majority of livelihoods (water scarcity and land degradation) as well as the drivers of long-term ecosystem degradation and loss of resilience. Considering cost, experience with adoption rates, and scalability, mobilizing additional support for the widespread adoption of sustainable land and water management practices represents a practical means to boost resilience: they provide productivity gains at household level and in most cases contribute to positive externalities at landscape level. Restoring degraded drylands by scaling up sustainable land and water management practices, both on and off farm in different landscapes (including rangelands, wetlands, and forests as well as farmland) can diversify production systems, increase rural incomes, and enhance the resilience of the most vulnerable households.

• *Integrated landscape management can protect investments in sustainable land management.* There is sufficient evidence that landscape interventions can help to manage rainfall runoff, restore groundwater tables, regenerate forests and pastures, and increase the density of trees on farms to increase biomass availability, and also boost wildlife numbers and biodiversity benefits. Investments by farmers, herders, and other land users in improved and more sustainable land management practices need to be safeguarded and conflicts need to be managed. Farmers and herders cannot be successful in adopting improved land and water management practices to restore landscape productivity and resilience if the drivers of land degradation such as uncontrolled grazing, cutting of trees, wildfires, land clearing, and other behaviors do not change. Integrated landscape management can support a process of bottom-up behavior change, reinforced by top-down policy and institutional reforms.

• *Integrated landscape management can move beyond safeguarding restoration efforts.* New integrated landscape management programs can do more and go beyond the restoration of degraded drylands and provide a framework for addressing the critical constraints to sustainable development at the landscape level. For example, consultative and planning processes and the institutional mechanisms developed to implement integrated landscape management interventions can also help to address marketing and transport bottlenecks, and support cost-effective investments in infrastructure development (for example, access roads, communications, capacity-building services) that contribute to further increases in the resilience and well-being of rural communities.

Integrated Landscape Management in Practice

Building on the existing work of EcoAgriculture Partners, the preliminary findings from a Landscapes for People, Food and Nature review of integrated landscape management initiatives in Africa and Latin America, and other implementations of a landscape approach to achieve sustainable development objectives, this section provides a brief overview of integrated landscape management examples and some lessons learned. Table 2.4 highlights some of the differences and common elements among these examples. Chapter 5 elaborates on these examples and takes an in-depth look at three case studies in Africa.

Integrated Landscape Approaches for Africa's Drylands • http://dx.doi.org/10.1596/978-1-4648-0826-5

The largest effort to date to catalogue integrated landscape management initiatives is being undertaken by the Landscapes for People, Food and Nature initiative, where researchers are conducting continent-by-continent surveys and interviews of landscape initiatives to investigate what is working and what is not. The Landscapes for People, Food and Nature initiative recently released its Africa continental review (Milder et al. 2014), and its Latin America review is currently being finalized (Estrada-Carmona et al. 2014). Preliminary results from the Latin America review found 104 examples of integrated landscape management initiatives in 21 countries. The Africa review found 33 examples in Sub-Saharan Africa in Kenya, Ethiopia, South Africa, Democratic Republic of Congo, and Uganda. Some of the major findings from these two reports include the following:

- Integrated landscape management initiatives represent a change in development approach. They represent an effort to leverage and scale up previous efforts in agricultural development and conservation through an intentional effort to use multi-stakeholder, multi-scale frameworks to plan and implement activities across an entire landscape in a more coordinated fashion.
- Integrated landscape management in Africa is not as widespread or well developed as in Latin America. However, the use of integrated landscape management in Africa is rising exponentially as there has been a dramatic increase in uptake over the past five years.
- The landscape initiatives were, in the majority, multi-sectoral. On average, more than three sectors were involved in each initiative in Africa, with natural resources/environment and agriculture being the most frequently cited.
- The private sector is largely missing from landscape initiatives in Africa, but is more active in Latin America.
- A wide variety of stakeholders are participating in landscape initiatives across Africa and Latin America.
- Based on the qualitative evidence base, there appears to be added value in using multi-objective integrated landscape management.
- Results suggest that integrated landscape management investments have contributed to the formation of new coordinating bodies, awareness of inter-sectoral linkages and co-dependencies, and increased institutional capacity, often within a few years.
- Temporary or unsustainable funding, unstable governance, and ineffective monitoring, evaluation, and implementation stand to undermine the success of integrated landscape management initiatives.

Other efforts have been undertaken to identify landscape approaches, although not necessarily with that terminology. Table 2.4 provides a sample of integrated landscape management efforts from selected regions around the world. The table demonstrates that projects have incorporated several if not most integrated landscape components and achieved both farm- and landscape-level benefits.

Table 2.4 Integrated Landscape Management Initiatives in Practice: Examples of Farm and Landscape-Level Benefits

Landscape, country	Landscape challenge	Main activities	Documented farm-level benefits	Documented landscape-level benefits
Africa				
Banikoara District **Benin**	Resolve conflict between pastoralists and cotton farmers.	Creation of livestock corridors through agricultural fields; "social fencing"; local agreements between farmers and pastoralists.	Increased livestock productivity from access to pasture and watering points; indirect benefit to cotton production through supply of manure and animal traction; livestock secondary source of income for farmers.	Reduced conflict along corridors.
Selected watersheds in Soum, Sanma-tenga, Kourit-tenga and Kompienga Province **Burkina Faso**	Biodiversity loss, ecosystem degradation, pressure on forests from agriculture expansion and unsustainable farming practices.	Introduce a landscape dimension and an Integrated Ecosystem Management (IEM) approach to local development planning; build local and institutional capacity for integrated ecosystem management; establish a Local Investment Fund (LIF) for Integrated Ecosystem Management (Sahel Integrated Lowland Ecosystem Management Project—SILEM).	Increased household incomes and adoption of new household-level technologies (improved soil and water management practices).	Adoption of collective technologies at inter-village level (collective river bank protection; animal route delimitation) based on IEM plans; positive trend in plant and insect diversity, and soil organic content (but the project's time period is too short to establish a robust conclusion).
Humbo **Ethiopia**	Restore highly degraded areas through forest landscape restoration.	The project is restoring 2,728 hectares of native forest through farmer assisted natural regeneration. The United Nation's Clean Development Mechanism has issued 573,000 carbon credits which were purchased by the World Bank's BioCarbon Fund.	Farmers are expected to receive revenue benefits from carbon sequestration payments and selling of wood products. Restoration is increasing availability of grasses that can be used as fodder and for sale.	Reduced erosion, improvements in soil health, protected groundwater supplies, carbon sequestration, and restored wildlife habitat.
Selected watersheds in Tigray **Ethiopia**	High population growth rates, frequent droughts and erratic rainfall, and poor access to sustainable agricultural technologies and market access left communities dependent on food aid and disaster.	Watershed restoration programs to complement food and cash for work programs that incentivized farmers into implementing soil and water conservation and rainwater harvesting interventions and collaborating with foresters and other communities.	Improvements in water table and soil nutrient levels and subsequently improvements in local agricultural production; reductions in soil erosion; livelihoods diversification.	Improvements in regional and district-level soil and water quality and quantity.

table continues next page

Table 2.4 Integrated Landscape Management Initiatives in Practice: Examples of Farm and Landscape-Level Benefits (continued)

Landscape, country	Landscape challenge	Main activities	Documented farm-level benefits	Documented landscape-level benefits
Two biological corridors **Ghana**	Land degradation, deforestation, and loss of biodiversity (major drivers were unsustainable harvesting levels in the savanna, inappropriate farming practices and annual wildfires).	Improve biodiversity and protected area (wildlife corridor planning); work in protected areas, forest reserves, and agricultural lands, adopting principles of a "landscape" approach to conservation and natural resource management.	Sustainable livelihood diversification (adoption of new household level technologies including mangoes, honey, low-tillage practices; community-based enterprise development linked to medicinal plants and ecotourism).	Established two biological corridors between Ghana and Burkina Faso to connect the two ecological blocks permitting wildlife movement (but impact on wildlife numbers not documented in evaluation report).
Kericho **Kenya**	Produce high yields of tea for international trade in area of highly threatened forest cover and biodiversity.	Unilever, an international food company, promoted sustainable agricultural practices among tea producers and other stakeholders, agricultural extension, tree-planting for fuelwood.	Committed to using renewable resources whenever possible, minimizing adverse effects of production on ecological health, and reducing resource inputs as much as possible, without reducing tea yields or quality; 10–15 percent increase in tea revenues.	Less pressure on forest resource and biodiversity; micro-climate effects.
Communal wildlife conservancies **Namibia**	Loss of habitat, decline in wildlife populations, limited livelihoods opportunities for local communities, human-wildlife conflicts.	New law gave local communities right to delineate a geographic area and create conservancies; these can be managed for multiple benefits including wildlife (for example, ecotourism, trophy hunting, meat), livestock, and, where possible, crop production. Local plan establishing priorities for conservancy; implementation of plan is monitored by conservancy members.	Households in conservancies increased their income from new economic opportunities.	Increased wildlife numbers.
Sahel parklands **Niger**	High population growth rates, frequent droughts, erratic rainfall, high incidence of rural poverty, frequent conflicts over resource use, especially between farmers and herders.	Reform of Forest Code, land tenure reforms and popularization of the Rural Code, promotion of decentralized natural resource management as an integral part of rural development strategies, farmer-to-farmer visits to promote scaling up of agroforestry, rainwater harvesting, micro-dosing, and strengthening of local land management institutions.	Rehabilitation of degraded lands, increased crop yields, expansion of dry season irrigated gardens, reduced food insecurity, diversification of household incomes, and increased resilience.	Landscape-level "re-greening" across millions of hectares, increased recharge of shallow aquifers, replenishment of fuelwood stocks in rural areas (micro-climate effects/wind speed/temperature).

table continues next page

Table 2.4 Integrated Landscape Management Initiatives in Practice: Examples of Farm and Landscape-Level Benefits *(continued)*

Landscape, country	Landscape challenge	Main activities	Documented farm-level benefits	Documented landscape-level benefits
Various landscapes **Sahel, Africa**	Address land degradation and climate change risks in the Sahel at the regional level by linking national-level efforts for restoration across Burkina Faso, Chad, Djibouti, Eritrea, Ethiopia, Mali, Mauritania, Niger, Nigeria, Senegal, and Sudan.	Expand sustainable land and water management interventions and technologies; improve land-use planning at watershed or local scales; improve and expand information base to enhance capacity of governments to address land degradation (Great Green Wall Initiative).	Expected impacts: improvements in crop production, tree density, soil fertility, erosion prevention, fodder availability, and strengthened resilience to climate change.	
Various landscapes (for example, Biodiversity and Wine Initiative, Western Cape) **South Africa**	Ecosystem degradation, species loss, and loss of natural habitat in production landscapes (for example, farming and grazing areas). Unconnected protected areas alone will not conserve a representative sample of biodiversity, maintain major ecosystem services, and safeguard ecosystem resilience. A landscape approach is needed to reduce vulnerability to climate change risks (both to biodiversity and provision of ecosystem services) or to exploit opportunities to enhance resilience.	Slow the loss of natural habitat in production landscapes (farming areas) and encourage biodiversity-compatible production methods without undermining farming livelihoods; increase critical biodiversity areas (for example, wildlife corridors) and ecological support areas (for example, areas providing freshwater and other ecosystem services); establish systematic biodiversity planning at various scales that takes land and resource use into account; establish land-use guidelines and biodiversity stewardship agreements with public and private landowners (multistakeholder partnerships).	Clearly identifiable branding of wines from participant wineries and marketing support by industry group providing unique selling point. No documented increase in sales or price premium.	Biodiversity guidelines are embedded in the South African wine industry; areas under conservation exceed the total area of planted vineyards in the Western Cape. As of 2010, over 204,000 hectares were under better land management including over 126,000 hectares of natural lands (of which almost 20,000 hectares are formally conserved).

table continues next page

41

Table 2.4 Integrated Landscape Management Initiatives in Practice: Examples of Farm and Landscape-Level Benefits *(continued)*

Landscape, country	Landscape challenge	Main activities	Documented farm-level benefits	Documented landscape-level benefits
Luangwa Valley **Zambia**	Unsustainable production of cash crops had depleted soil nutrients and increased farmers' reliance on poaching of wild animals for subsistence.	Promoting food security through increased training on technology and land-use practices to promote agricultural production and reduce poaching; zero-tillage training in exchange for traps and snares.	Crop production increased through use of zero tillage, cover crops, crop rotation and natural fertilizers; 15 percent increase household food security.	Less soil erosion; lower pressure on wildlife and protected areas.
Asia				
Loess Plateau **China**	High population growth rates and overgrazing and overuse had led to high levels of erosion, declining food supply, and poverty.	Loess Plateau landscape restoration through reforesting slope areas and leveling land to produce high-yielding crops.	More efficient crop production on terraces, diversification of agriculture, and livestock production. Per capita grain output increased from 365 kg to 591 kg/year. Income increased from US$70 to $200 per person per year.	Improved ecological conditions upstream. Improved ecological conditions downstream: reduced sediment load in tributaries of the Yellow River.
Rajasthan **India**	Environmental degradation and drought left dryland farming communities unable to meet water needs.	Collective community investments to re-establish and manage johads, traditional large-scale water harvesting structures.	Increased access to water for irrigation enabling 23 communities to have an additional crop-growing season and increased livestock production.	Improvements in regional soil and water quality and quantity.
Wanggameti, Sumba Island, Nusa Tenggara province **Indonesia**	Boundary and tenure disputes, livestock grazing conflicts, fire management and illegal logging in and around protected forest and nature reserve; poverty of smallholder farm families.	Established over 5,000 family forests (on-farm); promoted soil and water conservation; improved soil fertility through integrated farms (including livestock) under semi-arid conditions; multi-stakeholder Wanggameti Conservation Area Forum.	Increased agricultural and livestock productivity and improved soil and water retention for production on hillside farms for more than 3,400 rural households across 22 communities, reaching 17,400 beneficiaries; established over 5,000 family forests as sources of fuelwood, fodder, timber and non-timber products.	Reduction of negative downstream impacts from poor soil and water management upstream; reduction in conflicts.

table continues next page

Table 2.4 Integrated Landscape Management Initiatives in Practice: Examples of Farm and Landscape-Level Benefits *(continued)*

Landscape, country	Landscape challenge	Main activities	Documented farm-level benefits	Documented landscape-level benefits
Cebu watershed **Philippines**	Reverse the environmental degradation of the watershed and improve rural livelihoods for smallholder farm families.	Established soil and water conservation and integrated farms (including livestock); participatory innovation development for small-scale, hillside agricultural production.	Increased agricultural and livestock productivity on hillside farms for more than 1,500 rural households across 18 communities, reaching 8,100 beneficiaries; efficiency on-farm water management decreased yield crop fluctuations due to erratic weather patterns.	Reduction of negative downstream impacts from poor soil and water management upstream.
Kalinga **Philippines**	Maintain local livelihoods though irrigated rice cultivation and sustainable harvesting of wild animals.	Conservation and rehabilitation of integrated rice terraces; improved forest management practices; restored effective indigenous practices.	Over 150 hectares of rehabilitated rice terraces where fish and vegetable production also occur.	Reduction of negative downstream impacts from poor soil and water management upstream.
Americas				
Talamanca **Costa Rica**	Improve rural livelihoods of indigenous cocoa farming communities in ways that protect high biodiversity in the Mesoamerican Biological Corridor.	A regional organic small farmers cooperative promoted sustainable agriculture and forestry, ecotourism, and biodiversity conservation enterprises among members; organic certification and marketing of high-value crops; community-led biodiversity monitoring.	1,500 farmers using improved eco-friendly production practices, and increased competitiveness; increases of 15–60 percent in small-farmer revenue due to crop diversification and certification premiums. Increases of 15–60 percent in small-farmer revenue; US$70 to $200 per person per year.	
Matiguas **Nicaragua**	Slow the rapid expansion of livestock production to remaining forested areas.	Promoted silvo-pastoralism (incorporating tree cover into pasture lands) and payments for ecosystem services to promote biodiversity conservation and carbon sequestration.	Payment for ecosystem services income.	Silvopastoral practices adopted in 24 percent of the total project area; the total area of degraded land fell by two-thirds.

table continues next page

Table 2.4 Integrated Landscape Management Initiatives in Practice: Examples of Farm and Landscape-Level Benefits (continued)

Landscape, country	Landscape challenge	Main activities	Documented farm-level benefits	Documented landscape-level benefits
Various landscapes **Colombia,** **Costa Rica,** and **Nicaragua**	Address problems associated with pasture-based cattle farming including loss of natural habitat and biodiversity.	Promote a market-based approach by compensating land users who adopted silvo-pastoral practices in degraded pasture areas.	Demonstrated improvements in farm-level water infiltration, soil productivity, and reduction of fossil fuel dependence.	Restoration of pasture and increase in tree cover and wildlife habitat.
Malpai Borderlands, New Mexico and Arizona **USA**	Loss of grasslands, increased presence of poor quality species for grazing due to natural fire suppression, loss of biodiversity, reduced rancher income.	Reintroduced fire as a grassland management tool; reintroduced the keystone species, the black-tailed prairie dog; over half of the Malpai Borderlands placed in easements.	69,000 acres have been burned to reduce the presence of woody plants and increasing growth of economically useful perennial grasses. "Grass banking" is available to farmers experiencing severe drought.	Less biomass available to fuel more powerful wildfires; biodiversity benefits.

Sources: Conservation South Africa, 2011; Cadman et al. 2010; EcoAgriculture Partners 2012b; EcoAgriculture Partners 2013; GEF 2014; Pagiola and Arcenas 2013; World Bank 2009; World Bank 2011; World Bank 2012; Winterbottom et al. 2013.

CHAPTER 3

Role of Public and Private Stakeholders in Integrated Landscape Management

Landscapes consist of both interacting ecosystems and actors from multiple sectors and backgrounds. Stakeholders will naturally vary in how they utilize or perceive an ecosystem and its goods and services, and consequently, their motives for participating in a landscape approach will vary. Reconciling differences between stakeholder perspectives and motivations should be an important component of integrated landscape management to minimize costs and maximize social welfare gains and other benefits.

This chapter aims to analyze the role of public and private stakeholders within a landscape, identify winners and losers, and identify approaches for reconciling differences between resource users and society. With regards to natural resources, the Food and Agriculture Organization of the United Nations (2000) defines stakeholders as, "people who possess an economic, cultural, or political interest in, or influence over, a resource." Stakeholders of integrated landscape management are divided into three categories based on Karl (2000):

- *Primary stakeholders* are those that are targeted for participation in integrated landscape management because they both rely on and impact the health of the landscape and will be directly engaged in and impacted by integrated landscape management interventions.
- *Secondary stakeholders* are those who implement, administer, manage, or oversee integrated landscape management.
- *External stakeholders* are those not formally participating in integrated landscape management but have a stake in or can influence its outcomes in that they provide funding for or interact with primary and secondary stakeholders.

The following stakeholder analysis first identifies common stakeholders from integrated landscape management initiatives in Africa based on evidence from the case studies in Ethiopia, Kenya, and Niger, and a literature review.

Specific attention is paid to stakeholder motivations and interests in a system or a program identifying who stands to gain and forgo benefits from integrated landscape management. Stakeholder roles are described in the context of natural resource or ecosystem management, that is, the common interest that binds actors of integrated landscape management efforts. A discussion on the economic rationale of participation for each major stakeholder group follows, and the chapter concludes with a discussion of policy options for reconciling various stakeholder rationales.

Identification of Key Stakeholders and Rationales for Participation

A broad level of key actors for integrated landscape management in Africa and Latin America can be identified, based on the Landscapes for People Food and Nature initiative regional reviews (Estrada-Carmona et al. 2014; Milder et al. 2014). Some common stakeholder groups in both Africa and Latin America are:

1. Producer groups: farmers, livestock keepers, and agro-pastoralists
2. Community groups: agriculture and livestock cooperatives, water users' associations, community or village committees, women's self-help groups
3. Marginalized groups: could include indigenous groups, landless people, hunter-gatherers, and women
4. Government agencies: local up to national-level ministries and agencies
5. Donors
6. NGOs: both local and international
7. Education and research organizations: research institutes and academia
8. Agribusiness and industry

Milder et al. (2014) provide further insights into the stakeholder situation in Africa, stating that:

- Government entities were involved in the design or implementation of at least 95 percent of the identified integrated landscape management initiatives. Most were local- and district-level government entities.
- NGOs from both within and outside integrated landscape management initiatives were active in the majority of initiatives.
- Producer groups were active in 83 percent of initiatives.
- Women were participating in 57 percent of the initiatives.
- The majority of initiatives included participation by actors from across at least three sectors (the most common sectors involved were environment, agriculture, and natural resources).
- Agribusiness and industry were largely absent from African integrated landscape management initiatives.

Results from the case studies in chapter 5 confirm these findings. Table 3.1 identifies these key stakeholders.

Table 3.1 Key Stakeholders in Ethiopia, Kenya, and Niger

Stakeholders	Tigray, Ethiopia	Mt. Kenya, Kenya	Maradi and Zinder, Niger
Primary	• Smallholder farmers • Livestock herders and pastoralists • Women • Community groups	• Smallholder farmers • Community groups governing the use of forests • Pastoralists	• Smallholder farmers • Women • Community groups governing local land and resource use • Wood cutters/sellers • Pastoralists
Secondary	• Government agencies: Ministry of Agriculture; regional, zonal and *woreda* bureaus; community-level development agents • Common Interest Groups • NGOs	• International donors (IFAD, GEF) • Various water sector organizations (WRUAs, WRMA) • Common Interest Groups (CIGs) with 20–30 members • Community Forests Associations (CFAs) • Government agencies at national, county, and other scales • Private sector service providers with a comparative advantage	• Government of Niger (Forest Service, Agricultural Service, District government officials) • International and national NGOs (SIM, CARE, World Vision, Peace Corps) • African Regreening Initiative • National education and research organizations (university researchers) • Donors for rural development projects
External	• International and national NGOs (EcoAgriculture Partners) • Non-targeted households • Agricultural and environmental education/research groups • International organizations (USAID, GTZ, GIZ, EU, World Bank, United Nations World Food Programme) • Agribusiness	• Downstream water users (hydropower, irrigation, municipal water supply) • Other private sector actors operating in the region • Other national and international development cooperation actors operating in the region	• International relief and development assistance agencies • International education and research organizations

Note: CARE = Cooperative for Assistance and Relief Everywhere; GEF = Global Environment Facility; GTZ/GIZ = Gesellschaft für Technische Zusammenarbeit/Gesellschaft für Internationale Zusammenarbeit; IFAD = International Fund for Agriculture Development; SIM = Serving in Mission; USAID = U.S. Agency for International Development; WRMA = Water Resource Management Authority; WRUAs = Water Resource Users' Associations.

In all three country cases, the primary stakeholders are mostly smallholder farmers and herders. They are also the primary beneficiaries of integrated landscape management and their behavior ultimately impacts the outcomes of this approach and as such, they are generally the most important stakeholder group to target when it comes to improving resilience.

Beneficiaries will vary, however, depending on the target area and landscape objectives. Major stakeholder groups of integrated landscape management initiatives in Africa are described more broadly below, along with their function and roles and rationales for participation. The section concludes with a description in terms of who stands to gain or lose from integrated landscape management efforts.

Producer groups are critical to the success of integrated landscape management as they are generally the targeted beneficiaries, their behavior is intricately linked to landscape health, and they provide a wealth of indigenous/traditional knowledge. Crop cultivation is the primary source of livelihood in African dry-lands (IUCN 2010) and smallholder farmers and pastoralists are the primary stakeholders in our African case studies. Farmers' activities directly impact the quality of soil, water, and land. Integrated landscape management approaches operating in India and Africa have found it necessary to compensate farmers for short-term losses associated with implementing sustainable land management interventions which require them to forgo migratory labor income as well as some agricultural production and livestock grazing. In Ethiopia, farmers are compensated through food and cash payments for work through programs like the Productive Safety Net Program (PSNP) and the Managing Environmental Resources to Enable Transition (MERET) program. These payments improve food security of the poorest households. In India, farmers participating in water-shed development approaches have been compensated by wages for labor to support sustainable land management interventions and financing for revolving credit maintenance funds (Gray and Srinidhi 2013).

Other producer groups include agro-pastoralists and pastoralists and they act as both primary and secondary stakeholders. IUCN (2010) estimates that pasto-ralism is the second most widespread livelihood in eastern and southern Africa region's drylands, with over 40 million pastoralists and agro-pastoralists within the Greater Horn of Africa region alone. Livestock systems can potentially improve ecosystem quality as it is important for nutrient cycling, seed dispersal and germination, and vegetation management. Livestock products like manure can also serve as an input to other sectors like agriculture. Conversely, livestock systems can result in degradation and conflict with agricultural producers if not managed effectively. These producer groups are also short-term focused and their motivations can be at odds with other producer groups like farmers, especially if integrated landscape management requires grazing bans.

Community groups include organized groups and cooperatives operating within a landscape to achieve goals like improving the local-level governance of natural resources and ecosystem services, influencing policy and landscape approach development, providing a voice to marginalized communities in meetings, and aggregating sellers of similar products like crops, livestock, or for-est products to improve market access and profitability. Community-based organizations need to be empowered to negotiate, adopt, and enforce rules governing access and use of natural resources across rural landscapes, particu-larly in the case of common pool resources; this function is often critically important to the long-term success of improved land and water management interventions both on and off farm. Well-functioning community groups can help to identify short- and long-term solutions related to community needs. Community groups might also seek to mobilize community resources to pro-tect livelihoods and cultural identity. Examples of community groups include water users' associations, community interest groups, community forestry asso-ciations, livestock and crop cooperatives, traditional and religious groups or authorities, and women's self-help/microfinance groups. The different groups may have conflicting interests, but their existence means that they can communicate and negotiate with one another.

Marginalized groups can include a variety of individuals including hunter-gatherers, women, and indigenous communities. Like producer groups, their participation in initiatives may be motivated by equity considerations and appropriate distribution of benefits produced by interventions. Marginalized groups are often not involved in decision making and can be exposed to downstream or off-site negative externalities from integrated landscape management. For example, pastoralists and hunter-gatherers in East Africa and the Horn of Africa are often politically weak and pastoral institutions are often not recognized by law (Flintan, Behnke, and Neely 2013). The deliberate inclusion of otherwise marginalized groups has the potential to positively impact integrated landscape management. For example, the training and education of women in farming and microfinance can significantly increase the impact of interventions aimed at boosting agricultural production and can help support new business development.

Government agencies at local, national, and regional scales are usually secondary stakeholders as they provide financial, technical, administrative, and policy support. Examples include Ministries of Agriculture, Environment, Finance, Natural Resources, and Rural Development. This stakeholder category also includes regulatory agencies such as Environmental and Land Protection authorities at different levels. Government agencies decide the type of sustainable land management interventions as well as the mechanisms for implementation. They can have a variety of roles and often multiple agencies and levels of government are involved. In Ethiopia for example, the federal Ministry of Agriculture provides funding and implementation assistance, and regional, zonal, and *woreda* bureaus assist in project implementation and technical assistance (Zeleke et al. 2006). Government agencies' motivations for participation include improving domestic food security and reducing conflicts over natural resources, but also responding to domestic and international demands and agreements for drylands restoration (for example, many African governments actively participate in the United Nations Conference to Combat Desertification). In landscape approaches identified in Africa, government involvement also varies depending on whether the approach is more bottom-up (for example, Niger) or top-down (for example, Ethiopia).

Public and private donors and financiers can provide funding to landscape approaches in the form of grants, donations, investments, loans, and a variety of other financial instruments. Donors include bilateral and multilateral funding agencies (for example, World Bank, USAID, European Commission Global Climate Change Alliance), regional development banks (for example, African Development Bank), private investors (for example, carbon finance, offset brokers, insurance companies, commercial banks, conservation funds), and some public-private partnerships. Some donors provide financial support but are not heavily involved in participation, while others provide both financial support and work with other stakeholders on implementation and administration. Bilateral and multilateral donors generally have broader, longer-term motivations for funding landscape approaches, such as providing humanitarian assistance, improving relations with other countries, or meeting financial obligations established by national priorities or international agreements. Private investors can have shorter-term motivations and tend to make sector-specific investments (Shames, Clarvis, and Kissinger 2013). Shames, Clarvis, and Kissinger (2013) state that public and private investors face three key challenges that can constrain their investment in integrated landscape management:

• First, integrated landscape management requires integration across sectors and stakeholders, but if appropriate coordination platforms are not in place, integrated investments can be difficult to make. Practice shows that donors/financiers are often highly siloed, making landscape coordination difficult.
• A second challenge is that donors and financiers can view integrated landscape management as highly risky with uncertain economic returns compared to conventional sectoral approaches, which can discourage their investments in integrated landscape management efforts.
• A third challenge for donors and financiers is that integrated landscape management initiatives might have longer payback periods, which make them less attractive for investment.

Zeleke et al. (2006) identify a fourth challenge, stating that donors have their own financing procedures and reporting requirements which can create problems for integrated landscape management in terms of the efficient use of support (Zeleke et al. 2006).

Agribusiness and industry: Milder et al. (2014) and Estrada-Carmona et al. (2014) found that private sector activity in landscape initiatives in Africa is largely absent. Agribusinesses are mostly external stakeholders and can include farm machinery, seed, and chemical manufacturers, as well as companies involved with distribution, sales, and marketing of agricultural and forestry products. It is likely that the private sector's interest in landscape approaches is also more short-term focused as industrial operations are largely profit-driven. Agribusiness and industry tend to be more interested in high-potential agro-ecological zones where intensification is possible.

Nongovernmental organizations are active participants in integrated landscape management for a variety of reasons including equity, environmental rehabilitation, biodiversity protection, human welfare and health, and climate change. As such, their motivations for participation can either be short- or long-term focused, and can vary from improving livelihoods at the local scale to addressing global challenges. NGOs have actively participated in landscape approaches in Ethiopia, Kenya, and Niger and their activities have ranged from providing technical expertise on landscape approach activities to government agencies, producer groups, and marginalized groups, to working on policy development. Zeleke et al. (2006) found that for sustainable land management activities in Ethiopia, NGOs have concentrated on highly degraded areas and their reach has often been limited to very small areas. Zeleke et al. (2006) also remark that in

Ethiopia, NGO interests have not always been directly aligned with the interests and needs of producer groups and government agencies. For example, while NGOs can experiment and innovate, they may lack financial resources and expertise for long-term monitoring and technical assistance provision needed to sustain and scale up integrated landscape management activities.

Education and research organizations include higher learning institutes (for example, agricultural universities and colleges, technical and vocational training), agricultural extension services, research institutes and centers, and environmental think-tanks. Education and research organizations operate at both local and international scales. Their motivations range from providing technical expertise on the ground to improve livelihoods and food security, to creating awareness around sustainable land management activities and sharing lessons learned from around the world. They may or may not be actively involved in implementation and may or may not be directly impacted by integrated landscape management.

Table 3.2 summarizes stakeholder roles and rationales for participation, and also identifies stakeholders whose interests might potentially conflict with other stakeholder groups. For example, agricultural producers' interests might conflict with livestock keepers' interests if a landscape approach requires a grazing ban or

Table 3.2 Example of Stakeholder Analysis Matrix for Sub-Saharan African Drylands

Stakeholder	Level of involvement	Why are they important?	Rationale for participation (interests)	Could conflict with interests of:
Producers	Primary or secondary	• Responsible for land management activities that may or may not be sustainable, and which can either lead to degradation or restoration • Crucial for implementation of interventions • Interests may conflict between producer groups and other stakeholders, which could result in ineffective implementation or long-term failure	• Improve their food security and financial stability • Improve access to water • Improve technical skills • Cultural site protection	• Marginalized groups • Other producer groups • NGOs • Community groups • Government agencies • Public and private donors • Agribusiness
Community groups	Primary or secondary	• Mobilize people for participation in integrated landscape management activities and policy participation • Can play a critical role in local-level environmental governance (negotiation, adoption and enforcement of rules related to access and use of natural resources)	• Improve access to markets • Improve community representation in policy and program development • Improve equality in terms of access to resources, assistance, and income • Ensure sustainable use of common pool and other resources	• Marginalized groups • Producer groups • NGOs • Government agencies • Public and private donors • Agribusiness

table continues next page

Table 3.2 Example of Stakeholder Analysis Matrix for Sub-Saharan African Drylands *(continued)*

Stakeholder	Level of involvement	Why are they important?	Rationale for participation (interests)	Could conflict with interests of:
Marginalized groups	Primary or secondary	• Dependent on natural resources for livelihoods • May be a source of conflict over resource rights if excluded and not represented in local governance • May need assistance with small business development and micro-credit	• Improve their access to natural resources • Secure their livelihoods and increase their income • Improve equality in terms of access to resources, assistance, and income • Improve knowledge and skills development • Cultural site protection	• Producer groups • Community groups • Government agencies • Public and private donors • NGOs • Agribusiness
Government agencies	Secondary or external	• Development of policies/strategies/regulations • Development of land-use policies and land administration • Financing • Capacity building • Technical guidance • Monitoring and evaluation (M&E)	• Improve the stability of an area; avoid or minimize conflicts • Advise on sustainable use and potential means to increase resource productivity • National food security and achievement of other national development goals	• Producer groups • Marginalized groups • Community groups • NGOs • Agribusiness
Public and private donors	External	• Financing • Project support • Technical support • Capacity building	• Humanitarian assistance • National and international ODA obligations • Private investment to support new technologies/programs	• Marginalized groups • NGOs • Producer groups • Community groups • Agribusiness
NGOs	Primary, secondary, or external	• Financing • Technical guidance • Advocacy • M&E • Project implementation	• Biodiversity conservation • Climate regulation • Food security • Livelihoods	• Producer groups • Agribusiness • Community groups • Marginalized groups
Education and research organizations	Primary, secondary, or external	• Support policy implementation • Technical guidance • Training and education of resource users • Policy support • Capacity building • Education/skills development	• Promote best management practices • Increase international awareness of local issues and solutions	• Marginalized groups • NGOs
Agribusiness	Primary or external	• Purchase goods for outside markets • Promote new plant varieties	• Improve profitability • Gain new business	• NGOs • Education and research organizations • Government agencies • Producer groups • Community groups • Marginalized groups

Sources: Grimble et al. 1995; Zeleke et al. 2006; Nsouli 2000.
Note: ODA = official development assistance.

livestock enclosures that require livestock keepers to use new routes or new sources of fodder. NGO interests might also conflict with those of producer groups like smallholder farmers due to varying objectives. For example, small-holder farmers living in impoverished dryland regions might be more interested in short-term income and food security gains, while NGOs might be more interested in long-term environmental objectives like restoring biodiversity and improving groundwater supplies. As a result, farmers might be more willing to adopt farming practices and sustainable land management measures that maximize their short-term income even if they are not conducive to achieving long-term landscape approach objectives. The next section explores policy options available to address differences in rationales and perspectives for participation in integrated landscape management initiatives.

Policy Actions to Address Differences in Rationales

Integrated landscape management can help to reconcile conflicts of interest among the various stakeholders and play a role in considering how benefits could be distributed amongst potential winners and losers to address inequality and perceived risks. Considerations such as discrepancies in short- versus long-term motivations and potential conflicts are an important component of stakeholder analysis that can help to identify policies. Based on a literature review this section discusses several policy options that can be adopted by both government and other implementing entities within a landscape to reconcile differences in rationales and perspectives and incentivize participation. These policy options are echoed in chapter 6.

Integrated landscape management can include elements that are top-down as well as bottom-up. In Sub-Saharan African drylands, government participation and policy approaches are critical factors. Based on a report by the Organisation for Economic Co-operation and Development (OECD 2013) on collective action, policy options can be identified (see figure 3.1) ranging from cooperation to coordination. These can be viewed as options for public policy approaches and for implementing integrated landscape management initiatives.

Policy approaches can be divided into three major categories based on figure 3.1: technical assistance and knowledge sharing, economic incentives, and regulatory instruments.

Figure 3.1 Policy Approaches for Collective Action Initiatives

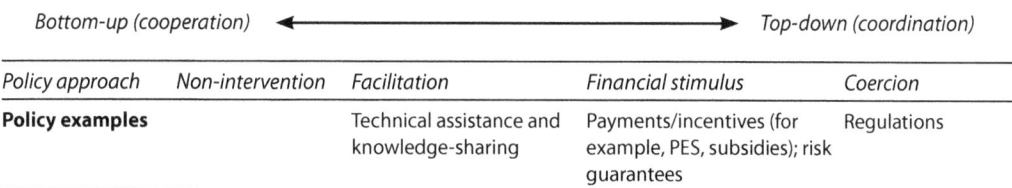

Bottom-up (cooperation) ←————————————————→ *Top-down (coordination)*

Policy approach	Non-intervention	Facilitation	Financial stimulus	Coercion
Policy examples		Technical assistance and knowledge-sharing	Payments/incentives (for example, PES, subsidies); risk guarantees	Regulations

Note: PES = payments for ecosystem services.

Integrated Landscape Approaches for Africa's Drylands • http://dx.doi.org/10.1596/978-1-4648-0826-5

Technical Assistance and Knowledge Sharing

Lack of information can be a major barrier towards effective adoption of integrated landscape management. Participation and multi-stakeholder involvement are critical components of landscape approaches as they help to ensure that communities acquire the knowledge and skills to maintain interventions over time and to spread awareness of the importance of regenerating ecosystems to provide long-term economic gains. Additionally, it is critical that primary, secondary, and external actors have information on how their respective roles will contribute to drylands restoration and how they will benefit over time. For example, the failure of Project Ethiopia 2488 (the precursor to the Managing Environmental Resources to Enable Transition project—see chapter 3) in the Tigray region of Ethiopia to meet food security goals is partially attributed to lack of maintenance of sustainable land management infrastructure, showing the importance of education and training for sustainable land management practices to ensure long-term success. Policies that can contribute to reducing information barriers and improving acceptance include:

- *Technical guidance documents*. Informational barriers can be a source of conflict between land user groups. Policies can focus on information dissemination, requiring guidance documents and meetings to help reduce information barriers and increase awareness of the benefits of landscape approaches. Technical guidance can be developed by government agencies, NGOs, and others, and can also be based on publicizing the achievements of farmer innovators.
- *Farmer-to-farmer learning*. Government agencies and NGOs can help facilitate cross-visits, farmer-to-farmer visits, peer-to-peer learning and other means of information sharing, communication, and outreach to help scale up interventions.
- *Creation of coordinating platforms and knowledge systems*. For governments, international organizations, NGOs, and others involved in drylands restoration, Flintan, Behnke, and Neely (2013) state that better information sharing is needed to improve coordination. NGOs, academic and research institutes, and others can be instrumental in creating virtual platforms, knowledge hubs and databases to share best practices for integrated landscape management and sustainable land management interventions, as well as knowledge of land use, tenure, and management systems, and relevant scientific information.
- *Develop tools to calculate risks and returns from integrated landscape management*. Government agencies, NGOs, and other promoters of integrated landscape management can support cross-sectoral investments and participation in integrated landscape management by improving awareness of risks and returns on investment for farmers, donors/financiers, and other relevant stakeholders. The World Resources Institute's Aqueduct tool, for example, helps private and public sector actors to understand business and development risks associated with water quantity and quality within a watershed.

Economic Incentives

Lack of financial and human capital can be major barriers towards integrated landscape management. Several options are available to help encourage investment and reconcile differences in perceptions of winners and losers from integrated landscape management initiatives. Below are some examples:

- *Fiscal instruments and public investments.* Although many sustainable land management and landscape-level interventions generate benefits relatively quickly as degradation is reversed and natural resource productivity is increased, depending on the types of interventions, the upfront costs of some landscape interventions may be perceived by specific stakeholders to be prohibitive. Resource users may not be willing to change their behaviors and land-use practices if they believe that the short-term costs will negatively impact their food and income security and that potential benefits will not be equally distributed. Budgetary measures including subsidies and taxes can be applied to help reduce costs or generate revenue to help offset initial costs and build stakeholder capacity. Additionally, food and cash payments for implementation of sustainable land management interventions through international donors and government agencies can help overcome short-term opportunity costs for producer groups. Public investments in road infrastructure can also help to enable access to markets and reduce transport costs (Global Mechanism and FAO 2009). Public investments in skills training and micro-credit opportunities can also help empower marginalized populations including women and youth (Flintan, Behnke, and Neely 2013).
- *Payments for ecosystem services (PES).* PES schemes are designed to internalize environmental externalities and provide incentives for investing in the provision of ecosystem services. PES schemes can help redistribute benefits achieved through improved ecosystem management and restoration by providing a mechanism for the beneficiaries of ecosystem services to pay for their provision. Payments can be designed to compensate landowners and others whose management decisions impact a landscape's quality and flow of ecosystem services. PES is a common incentive scheme with applications across the world. Several PES schemes have been developed, including:

1. Public payment schemes that commonly involve direct payments from, for example, a government agency to landowners to enhance ecosystem services;
2. Formal markets with platforms to allow trading of ecosystem service credits of, for example, carbon and water quality, between buyers (polluters) and sellers (landowners);
3. Self-organized private deals where individual beneficiaries of ecosystem services contract directly with landowners to enhance those services; and
4. Eco-labeling schemes that assure buyers that products they buy were sustainably made (Forest Trends, The Katoomba Group, and UNEP 2008).

- PES schemes can be mandatory or voluntary and can be designed with a good deal of flexibility in terms of how beneficiaries are charged and how landowners are compensated. PES schemes are being piloted across Sub-Saharan Africa. A report by the Global Mechanism of the UNCCD and FAO (2009) notes that some countries in the region have benefited from the carbon market and that the Clean Development Mechanism program is a potential source of funding. In Uganda, for example, PES schemes are being piloted to promote restoration of wetlands and to promote participation in carbon markets and conservation stewardship in two of Uganda's national parks. The Uganda-based NGO, EcoTrust, is also working to aggregate farmers and encourage planting of indigenous tree species and sell carbon credits via international carbon markets to the UK-based packaging company, Tetra Pak (Ruhweza, Biryahwaho, and Kalanzi 2008).
- *Innovative financing mechanisms that meet multiple objectives and reduce investment risk.* Incentives need to be created to encourage public and private investments at the landscape scale and across sectors, and to reduce risks associated with markets, politics, and natural disasters and ecosystem health (for example, severe droughts) which can lower returns on improvements in land productivity. Shames, Clarvis, and Kissinger (2013) provide a variety of recommendations for encouraging investment, including risk mitigation mechanisms (for example, private and public sector insurance, risk guarantee mechanisms), long-term bonds, and finance coordination mechanisms. These mechanisms are designed to protect investors from risks to expected returns on landscape investments and assist them with upfront investment costs. For example, USAID has a Development Credit Authority that provides credit guarantees agreements to private lenders to encourage loan provision to underserved borrowers in developing regions (USAID 2014). The guarantees allow lenders to take on additional risks as they know loans are more likely to be repaid.

Regulatory Instruments

Finally, government agencies can help reduce risks of integrated landscape management and encourage uptake through a variety of regulatory instruments that help to clarify stakeholder roles, reduce the potential for conflict, and overcome redundancies and inefficiencies in current efforts to address food security, natural resource restoration, and ecosystem health. Examples include:

- *Policies that clarify rights and responsibilities.* Individuals and communities whose land management activities influence a landscape should have clearly defined land and resource rights and land-use responsibilities. In the case of Niger, farmers participating in farmer-managed natural regeneration (FMNR) measures increased dramatically after they obtained management rights and *de facto* ownership over trees on their farms. Additionally, many sustainable land management activities require long-term investments and maintenance that require land tenure security to encourage adoption (Global Mechanism and FAO 2009). Clarification of rights and responsibilities can also help to reduce conflicts between resource user groups as it reduces uncertainty around access and use, and helps ensure that marginal groups are not disadvantaged.

- *Policies for land-use planning.* Careful land-use planning can be an effective strategy for discouraging activities that contribute to degradation (for example, deforestation) and for reducing development impacts. Land-use plans can identify conservation and restoration priorities, establish planning zones to limit development and degradation activities in priority areas, and identify stakeholders.
- *Biodiversity offsets.* Biodiversity offset schemes require developers whose activities result in a decline in biodiversity on a project site to purchase offset credits elsewhere (generally developed by private actors who create or expand habitat area). Offsets can be established as a government policy to ensure future development does not reduce biodiversity or important habitat.
- *Mechanisms for conflict resolution.* Conflict over resource rights and access can be a major barrier to effective implementation of landscape approaches. Governments can act to establish mechanisms for conflict resolution and compensation that are transparent (Global Mechanism and FAO 2009).
- *Strategies to harmonize and coordinate donor resources and drylands restoration programs and projects.* Stakeholders in integrated landscape management (governments, donors, NGOs, and international organizations) are operating in departmental, sectoral, or sustainable land management silos (Zeleke et al. 2006; Flintan, Behnke, and Neely 2013). Governments can act to promote systematic integration of these stakeholders by developing strategies and policies to improve communication to make landscape-level restoration initiatives more targeted and reduce programmatic redundancies (Zeleke et al. 2006).

Implications for Implementing Integrated Landscape Management

Based on findings from this stakeholder analysis and overview of policy approaches, the following are insights related to implementation considerations for integrated landscape management:

Prioritization of Stakeholders

Depending on the local context, some stakeholder groups should be prioritized for integrated landscape management to be successful, either because of their influence or impact on the landscape, or because they need special support via capacity building.

- Producer groups often represent a best first entry point for improving resilience in arid, semi-arid, and dry sub-humid zones because their livelihoods are most closely associated with changes to production systems and because of their indigenous knowledge. In countries such as Senegal with strong locally elected commune-level leadership, the best entry point might be strengthening local land-use planning through local government working closely with community groups. In other areas, the focus should be on small-scale farmers and associated community-based organizations because these represent the best bet for landscape-level restoration through sustainable land and water management. In Ethiopia, it was important to have producer groups on board early to implement sustainable land management interventions and improve their awareness. However, it was also important to provide education on what combination of sustainable land management interventions should occur and with what sequencing. This technical assistance was provided by NGOs and government agencies.

- Interactions between user groups should be considered to determine if interventions are needed to improve producer groups' (for example, between farmers and pastoralists) relationships to avoid conflict, which can lead to further land degradation. Historical relationship contexts should be taken into account.

- Creating producer user groups (for example, water resource users' associations or WRUAs) has been instrumental in giving producer groups a larger voice in development decisions and has helped to incentivize increased participation in drylands restoration projects.

- Mechanisms and platforms that encourage collective action and stakeholder collaboration and spread knowledge on lessons learned might be needed to encourage participation and reconcile stakeholder differences. These platforms can focus on financing, knowledge sharing on sustainable land management practices, and technical guidance for sustainable land management implementation and maintenance.

Enabling Conditions

Before drylands restoration efforts can be appropriately scaled or new integrated landscape management initiatives are created, some enabling conditions should be put in place or promoted to build long-term resilience.

- Government agencies are important for creating enabling conditions to support adoption and scaling up of sustainable land management practices, through initial policy and regulatory support. For example, in Ethiopia, the Environmental Policy of Ethiopia helped set the stage for integrated landscape management as it promoted sustainable resource use. Additionally, government support for community-based sustainable land management intervention through promotion of participatory integrated landscape management guidelines (for example, Ethiopia's Community Based Participatory Watershed Development guidelines) also can help create common entry points for participation across sectors and stakeholders. Finally, government support for more effective land-use planning can be instrumental in setting restoration and conservation priorities and bringing stakeholders together.

- In Niger and Ethiopia, government support for improved land tenure and increased security of resource access for producer groups has helped scale up sustainable land management practices and achieve landscape-level transformations that have generated significant benefits for many primary stakeholders.
- Within a country, a variety of actors working on different aspects of sustainable land management and integrated landscape management often do not seek to create synergies between their actions. Appropriate integration of the various policies, programs, and actors is key for promoting integrated landscape management and reducing upfront costs perceived by these stakeholders.

Who Wins and Who Loses: How to Encourage Multi-stakeholder Participation?

Integrated landscape management produces winners and losers, but mechanisms are available to deal with issues of political economy, incentivize participation, and compensate losers (see table 3.3). As will be discussed in chapter 4, farmers, pastoralists, and other producer groups will likely be major winners as resilience and productivity are restored. They could face, however, opportunity costs associated with implementing sustainable land management interventions to improve ecosystem health, such as lost income from labor migration and initial losses in livestock and agriculture income from exclosures. Government subsidies, investments, and payments (for example, wages and food for sustainable land management intervention and investments in social safety net programs) have been used successfully in India and Ethiopia to encourage farmer participation in drylands restoration. In some areas, market-based incentives such as Payments for Ecosystem Services schemes have helped to overcome the initial concerns of farmers and pastoralists about the feasibility of some proposed integrated landscape management interventions. In other countries, risk guarantees have helped encourage investment by private sector actors who view investments in ecosystems as too risky.

Table 3.3 Policy Options to Reconcile Differences between Winners and Losers

Technical assistance/ knowledge-sharing	Economic incentives	Regulatory instruments
Technical guidance documents that promote landscape principles	Fiscal instruments (for example, subsidies and taxes)	Policies to clarify rights and responsibilities
Peer learning programs	Food and cash for work payments	Land-use planning policies
Coordinating platforms and knowledge systems	Public infrastructure investments	Biodiversity offsets
Risk management/reduction tools	Payments for Ecosystem Services programs Risk mitigation mechanisms (for example, insurance, long-term bonds, credit guarantees)	Mechanisms for conflict resolution Policies to better harmonize and coordinate donor resources and drylands restoration programs

CHAPTER 4

Economic and Ecological Evidence on Integrated Landscape Management

This chapter provides economic and ecological evidence on the added value of integrated landscape management for drylands development in Sub-Saharan Africa in terms of increasing resilience and reducing vulnerability. In other words, the whole (integrated landscape management) is greater than the sum of its parts (individual sector-specific sustainable land management interventions). The chapter begins with a discussion of economic and ecological evidence from integrated landscape management initiatives across India, Latin America, and Africa based on a literature review of experiences identified by the Landscapes for People, Food and Nature initiative. This is followed by insights into economic valuation and evaluation challenges and limitations and how these factors influence the analysis and understanding of integrated landscape management.

A second goal of this chapter is to provide a theoretical discussion of the potential unique and additional costs and benefits of a landscape approach compared to sectoral approaches. It takes a closer look at integrated landscape management in Ethiopia, where integrated watershed management approaches have been active and studied since the early 2000s, providing an appropriate amount of time to review costs and benefits. Due to data limitations and challenges, the review looks at multiple landscape projects active in Ethiopia as opposed to assessing a single project. For example, many programs operate in the same area, and it is possible that households could have participated in more than one program, which would make attributing results to one program difficult. This chapter does not provide a thorough cost-benefit analysis but rather a discussion of Ethiopian landscape approaches in light of a cost-benefit framework developed as part of the theoretical discussion. Finally, it discusses the relevance of potential economic and ecological gains from improving resilience to shocks and stressors.

Review of Valuation Approaches and Challenges

To gain a better understanding of potential economic and ecological gains of landscape approaches compared to sectoral approaches, a literature review of integrated landscape management projects, programs, policies, and interventions was conducted, focusing on Asia, Latin America, and Africa. As mentioned in chapter 2, the term "integrated landscape management" encompasses a wide variety of programs and investment strategies.

Results from Asia

In India, the leading landscape approach for drylands restoration in rain-fed regions is termed "Watershed Development." Watershed Development began in the 1970s as a national strategy for improving crop production and reducing poverty in cultivable rain-fed areas that have experienced heavy degradation due to both human and climatic pressures including erratic rainfall and drought. These areas are seen as critical for improving India's food security as they represent almost 60 percent of the country's cultivable area and are also home to the majority of the country's rural poor (Government of India 2012). Watershed Development projects have evolved from being top-down, centralized, and focused on technical soil and water conservation interventions to being more integrated, bottom-up, and community-focused. They are now highly participatory and include social and ecosystem interventions (for example, micro-credit lending, community-based decision-making institutions, afforestation and reforestation) along with technical interventions aimed at controlling soil erosion and supporting sustainable land management (Kerr 2002). While the initial top-down projects that focused on soil and water conservation interventions improved crop production, net returns were low and benefits were unequally distributed between land-user groups, with tribal communities living in higher areas being disadvantaged. This is largely because they did not involve local communities, which resulted in insufficient maintenance and awareness of soil and water conservation interventions. The evolution of Watershed Development towards promoting integrated landscape management can be seen in the recent Twelfth Five Year Plan and national Watershed Development guidelines, the Common Guidelines for Watershed Development (that is, Common Neeranchal Guidelines), which both emphasize a holistic approach towards Watershed Development promoting equity, growth, and sustainability (Government of India 2011; Government of India 2012). For example, the Twelfth Five Year Plan states:

> Developing rain-fed areas requires pursuing three inter-connected goals simultaneously: enhancing current livelihoods for most people (equality), enhancing current carrying capacity (growth) and setting in motion regenerative processes to enhance future carrying capacity continuously (sustainability). This calls for re-shaping interactions between people and natural resources as well as those between nature's elements to produce multiple, long-lasting, synergetic effects rather than merely maximizing current production.

Additionally, these guidance documents call for better convergence of policy making, programming, budgeting, and implementing to make better use of resources and time.

At least two systematic meta-analyses of Watershed Development projects in India have been conducted (Kerr 2002; Joshi et al. 2005), as well as numerous economic valuations and project evaluations of Watershed Development projects (Sreedevi et al. 2006; Palanisami et al. 2009; Kale, Manekar, and Porey 2012). Joshi et al. (2005) conducted a meta-analysis of over 300 Watershed Development case studies. To determine the effectiveness and efficiency of these case studies, the authors assessed benefits related to employment generation, efficiency, and sustainability based on the following factors: geographic location, rainfall patterns, size of the watershed, focus of watershed on degraded land, implementing agency, people's participation, per capita income per region, activities or interventions performed, and soil types. The study found that benefits were greater in lower income areas, where people's participation was higher, and where rainfall ranged between 700 millimeters and 1,000 millimeters per year. The study also found a mean cost-benefit ratio of 2.14, indicating that benefits in Watershed Development amounted to more than double the initial investment, and a mean internal rate of return of 22 percent. Kerr (2002) conducted a meta-analysis of Watershed Development case studies in Andhra Pradesh and Maharashtra and found similar results. This study compared government-led projects that focused largely on technical improvements, NGO-led projects focused on social organization and integrated landscape management principles, and collaborative projects between government and NGO entities. Specifically, Kerr found that participatory projects performed better than top-down, technocratic approaches, but the projects that performed best were those that were participatory with sound technocratic input (that is, collaborative projects implemented by NGOs with government support or vice versa). Kerr states:

> The better performance of the more participatory projects seems to be related to the complex, site-specific livelihood systems prevalent in the study areas. These conditions call for a flexible approach and responsiveness to diverse, often unexpected situations. Blueprint approaches pursued by the technocratic, hierarchical organizations are poorly suited to such conditions. The NGO and NGO/government collaborative projects devoted time and resources to organizing communities to establish the locally-acceptable social arrangements for watershed interventions.

Kale, Manekar, and Porey (2012) and Palanisami et al. (2009) conducted economic valuation and impact studies of participatory Watershed Development projects. These studies found positive benefit-cost ratios indicating the projects generated positive economic returns. Kale, Manekar, and Porey (2012) found that the benefit-cost ratio increased from the first five years of project implementation to the second five years, indicating that economic returns increase over time as ecosystem health is restored. Sreedevi et al. (2006) conducted an impact evaluation of a participatory Watershed Development project in Gujarat and found that cropping intensity increased by 32 percent over eight years. Other on-site benefits included increased groundwater levels, crop diversification, improvements in fuel and fodder productivity, and improved income levels, literacy, and social development. Off-site benefits beyond the studied watershed included increased groundwater availability, reduced siltation and flooding, improved land productivity, and reduced distress migration.

In summary, case studies in India provide strong evidence as to the added value of integrated, participatory, and watershed-scale approaches. Key common findings include:

- The micro-watershed (around 1,000 hectares) is recognized as an effective unit of implementation. Many Watershed Development projects require participation by all community members in the watershed for sustainable land management measures. Studies have shown that participatory Watershed Development projects employing integrated landscape management principles have helped to protect natural resources, increase land productivity (especially crop, fodder, and forestry production), restore marginal lands, increase employment rates, and improve livelihoods of marginal populations including tribal groups and women (Kerr 2002; Joshi et al. 2005; Palanisami et al. 2009; Kale, Manekar, and Porey 2012).
- Several studies have shown that the net present value and/or economic and social benefits of Watershed Development projects increase over time, indicating a time lag between implementation of integrated landscape management interventions and improvements in ecosystem health and positive economic returns (Kale, Manekar, and Porey 2012).
- Multiple-stakeholder involvement, especially community involvement, in implementation is important for increasing economic and ecological benefits (Kerr 2002; Joshi et al. 2005; Palanisami et al. 2009). Additionally, bottom-up approaches that focus on capacity building and learning-by-doing approaches have been critical for success (Kerr 2002; Joshi et al. 2005; Palanisami et al. 2009).
- Watershed Development projects are able to produce on-site and off-site benefits by improving ecological conditions, namely, groundwater levels and improved soil quality and quantity (Sreedevi et al. 2006). Additionally, in many regions, the uptake of Watershed Development interventions has increased on non-treated sites as farmers witness improvements in land production on treated sites.

Another well-cited example of integrated landscape restoration is from the Loess Plateau in Northwest China. This area is home to over 50 million people and has a history of severe land degradation and erosion caused by unsustainable land-use practices including overuse and overgrazing. In 1994, the Government of the People's Republic of China with the assistance of the World Bank initiated the Loess Plateau Watershed Rehabilitation Project. This project was highly participatory and included a variety of social, environmental, and economic interventions such as: wide-scale terracing; agroforestry; development of roads to allow access of construction and farm equipment; enclosures for livestock; a grazing ban; creation of land-leasing options for farmers to improve ownership over interventions and allow farmers to profit from improved land productivity; and informal credit and loans to assist farmers in acquiring farm equipment and infrastructure (Mackedon 2012). Recent findings indicate that restoration efforts have almost tripled household incomes from US$70 to US$200 per person per year through improvements in and diversification of agricultural production (World Bank 2007). Other benefits generated by the project include: reduced sedimentation of waterways due to reduced erosion; improvements in land productivity due to increased regeneration of grasslands, trees, and shrub cover; female empowerment through increased work opportunities for women; improved production of food supplies; and income diversification due to reduction of labor input for farming (World Bank 2007).

Results from Latin America

In Latin America, the Landscapes for People, Food and Nature initiative regional review found 104 examples of landscape initiatives in 21 countries (Estrada-Carmona et al., *in review*). One landscape approach concept that is gaining traction is "Forest Landscape Restoration." Forest Landscape Restoration was first developed by the World Wildlife Fund and the International Union for the Conservation of Nature (IUCN) at a workshop in 2000 in response to the widespread failure of more traditional restoration approaches that focused on planting non-native tree species to generate a limited number of forest products and that ignored the root causes of forest degradation (Newton et al. 2012). Forest Landscape Restoration adopts many of the key integrated landscape management principles as it promotes participatory flexible processes, adaptive management and monitoring, restoring ecological processes at appropriate geographical scales to ensure ecosystem integrity, and multiple objectives and goals (for example, enhance human well-being, restore ecosystem services) (Newton et al. 2012). Forest Landscape Restoration has applications in many countries across Latin America. A study by Birch et al. (2010) estimated the cost-effectiveness of forest landscape restoration approaches by conducting spatial analyses of ecosystem service values from carbon storage, timber and non-timber forest products, tourism, and livestock production. They found forest landscape restoration initiatives to be cost-effective where passive restoration measures were used (for example, fencing and fire suppression). They also found that where active restoration was used (for example, native tree planting, fencing, and fire suppression), costs outweighed the benefits. The study only assessed market benefits however, and did not assess the co-benefits of forest restoration, which include improvements in human health and education, biodiversity and habitat improvements, and changes in water quality and quantity. Newton et al. (2012) found that forest landscape restoration initiatives can be cost-effective if the increased provision of these ecosystem services is taken into account.

Payments for Ecosystem Services and integrated silvo-pastoral approaches for ecosystem management are other common approaches found in Latin America, specifically Costa Rica, Nicaragua, and Colombia. These approaches aim to incentivize more sustainable land use, increased resource productivity, and awareness of ecosystem health.[1] For example, the Regional Integrated Silvopastoral Ecosystem Management Project (RISEMP) project in Colombia, Costa Rica, and Nicaragua, funded by the Global Environment Facility and the World Bank, aims to reverse land conversion for livestock production by incentivizing farmers and communities to adopt silvo-pastoral practices (for example, practices that combine livestock production with forestry). RISEMP piloted Payments for Ecosystem Services approaches in the three countries to help farmers overcome barriers to planting trees such as high upfront costs and time lag between investment returns. The projects were implemented by local NGOs and developed with support from the Livestock, Environment and Development initiative (LEAD) hosted by FAO. The Payments for Ecosystem Services schemes applied a differential payment method whereby farmers who adopted silvo-pastoral practices were paid (using GEF funds) based on changes in land use as estimated through an environmental service index (ESI). The ESI combined two indicators including biodiversity and carbon sequestration (Pagiola and Arcenas 2013). While these projects were implemented on a limited scale, the RISEMP approach exemplifies several landscape principles including participation by primary stakeholders and working across multiple sectors, namely forestry and livestock.

Results from Africa

In Africa, a variety of projects and programs that are active can be considered as promoting integrated landscape management. Milder et al. (2014), in their region-wide assessment of landscape approaches in Africa, provide an historical overview of relevant concepts in Africa which include "integrated conservation and development projects," "integrated rural development," and "integrated natural resources management." They found 33 examples of landscape approaches in Sub-Saharan Africa in Kenya, Ethiopia, South Africa, Democratic Republic of Congo, and Uganda.

German et al. (2012) examined results from the African Highlands initiative—an eco-regional research program designed to improve livelihoods and reduce natural resource degradation that can be classified as landscape—and found improvements in: crop yields; increased technology adoption; benefits associated with collective action including improved access to information, training and credit; improvements in housing, nutrition, and education enrollment; improvements in water quality and quantity provision; and increased awareness and rates of participation and cooperation.

A study on sustainable land and watershed management practices in the Blue Nile region of Ethiopia by Schmidt and Tadess (2012) found that the return on investment for sustainable land and water management infrastructure (including erosion mitigation and water conservation measures) could be positive over seven years (in terms of value of agricultural production and livestock holdings), indicating a time lag between implementation of interventions and improvements in land productivity and income. An interesting finding was that while positive gains in terms of agricultural production did not occur until the seventh year, the marginal benefits of sustainable land and water management interventions after the seventh year improved at an increasing rate. This seems to match findings for restoration projects in India where economic returns increased over time after restoring ecosystem health and functions.

Yitbarek, Belliethathan, and Fetene (2010) conducted a cost-benefit analysis of a watershed rehabilitation project in South Gondar, Ethiopia. The project used a participatory process and included revegetation of degraded lands, improved agricultural management practices, and creation of a land management plan. The study compared investment costs for project activities and compared them with benefits from carbon sequestration and increased forestry productivity resulting in biofuel and construction wood production. They found that benefits were greater than costs and a full return on investment was possible within four to seven years.

Limitations of the Literature

Despite these findings, the literature review revealed that the economic valuation and program evaluation on integrated landscape management is very limited especially in terms of off-site economic and ecological benefits and improvements in resilience. Barron and Noel (2011) conducted a review and synthesis of agricultural water management interventions in watersheds of 1–10,000 square kilometers in Asia, Sub-Saharan Africa, and Latin America. The study found very few benefits to cost evaluations at the full-project level and very few consistent and comprehensive analyses of benefit-cost ratios of watershed interventions and watershed management. EcoAgriculture Partners (2013b) states that hundreds of integrated landscape initiatives have been identified across the world but that little documentation exists as to their overall effectiveness and economic benefits. Sreevedi et al. (2006) also note that off-site impact studies for Watershed Development are lacking.

One reason for limited valuation efforts for integrated landscape management is that the concept is relatively new and is still constructively ambiguous, meaning different things to different people. Additionally, landscape initiatives in drylands are generally conducted in highly impoverished regions, where implementing organizations are often faced with limited capacity and difficulties in covering recurrent costs for medium- and long-term monitoring and evaluation. Two of the biggest causes of insufficient systematic project evaluation relate to a lack of project funding or support for monitoring and evaluation and a lack of technical expertise for conducting project evaluations and economic valuations. Barron and Noel (2011), Yitbarak, Belliethathan, and Fetene (2010), and Joshi et al. (2005) found a lack of consistency in data reporting, monitoring and evaluation, and project evaluations. Additionally, while project funders generally require project feasibility and evaluation reports, these reports are not often made publicly available, leading to a lack of transparency and understanding.

Another finding from the literature review is that economic valuations of integrated landscape approaches tend to focus on reporting productivity gains, namely improvements in agriculture, forestry, livestock, and fodder yields. These gains are easy to value as products have market prices. Studies neglect the potential co-benefits gained from landscape approaches in terms of provision of other public goods (that is, improvements in ecosystem service provision) and influence on positive and negative externalities that do not have market prices (for example, habitat, groundwater recharge, water quality improvement, improved nutrition, and human health and education) (Bach et al. 2011; Barron and Noel 2011). Newton et al. (2012) and Bach et al. (2011) note that there is a lack of understanding of potential project costs and benefits for any ecosystem-based restoration project due to lack of scientific evidence on how various interventions influence biophysical relationships and the provision of ecosystem services and human livelihoods. Conversely, there is also a lack of understanding in terms of how actions from various sectors are interacting to result in land restoration. As a result, there is a lack of understanding of the potential ecological and social gains, downstream or off-site externalities, the distribution of benefits across user groups, and change in behavior related to public goods. Finally, another limitation of the literature is a lack of analysis at multiple scales (Newton et al. 2012). Often, studies concentrate on demonstrating benefits at a farm or project scale—as a result, some of the economic benefits that might be generated due to greater ecosystem connectivity at a landscape scale are not captured. These benefits include those resulting from collective action and improved biophysical conditions.

Despite these challenges and data limitations, some interesting trends and lessons on how drylands restoration projects have evolved, their potential costs and benefits, and other economic valuation considerations from the literature can be identified. First, many countries are moving towards adopting integrated landscape management, often in association with increased interest in the restoration of degraded areas. This appears to indicate that the economic and ecological evidence to date has been sufficient to promote continued investment in programs that support integrated landscape management. For example, with regards to watershed management approaches similar to India's Watershed Development approach, the World Bank (Darghouth et al. 2008) states:

> National policies on watershed management have tended to develop in a pragmatic and iterative fashion, with early setbacks over engineering-dominated approaches being succeeded by tests of community-based approaches and by technology packages targeting sustainable changes in land and water use practices that brought profit to stakeholders. In several countries, including Brazil, China, India and Turkey, success in testing community-based approaches has led to adoption of broader policies for community-based watershed management.

Additionally, this indicates that countries are not seeing desired results from previous approaches, which have tended to be more top-down and focused on a few technologies in a specific sector. A second trend is that participatory projects involving multiple stakeholders, specifically farmers and livestock keepers, seemed to fare better in terms of generating economic, ecological, and social gains than top-down, centralized, sectoral approaches. Finally, the literature clearly identifies a need for greater evidence building by strengthening restoration planning, monitoring, and evaluation components to measure both upstream and downstream benefits, and farmer- versus landscape-level benefits. Studies point towards the importance of establishing clear, transparent, and quantitative baseline conditions and conducting iterative participatory monitoring and evaluation to track success and progress towards building resilience.

Economic Framework—Reviewing Unique Costs and Benefits of Integrated Landscape Management

To better understand the potential economic, ecological, and social gains from integrated landscape management, this section first identifies a framework to evaluate and compare landscape approaches with more top-down, sectoral approaches in terms of economic costs and benefits. This section provides some theoretical discussion on costs and benefits that should be assessed and compared. This is then followed by examples from landscape approaches implemented in Ethiopia, a region with severe land degradation and poverty and multiple programs focused on integrated watershed development to tackle food relief and sustainable land management. The section concludes with a discussion on how integrated landscape management is improving resilience and reducing vulnerability to climate change using the Ethiopia case study.

Cost Framework

In general, three main categories of costs are relevant to development projects: implementation costs, opportunity costs, and transaction costs (Ammann et al. 2013). Examples of each type of cost are provided in table 4.1 and below:

Table 4.1 Cost Evaluation Framework

Implementation costs
Capital expenditures on equipment and infrastructure
Annual operations and maintenance costs
Labor costs for administration and implementation
Opportunity costs
Foregone migratory labor income
Foregone income from previous (often) unsustainable activities (for example, logging, fuelwood collection, unsustainable agricultural practices, overgrazing of animals, etc.)
Transaction costs
Search costs: identifying program participants, identifying funding sources, etc.
Bargaining cost: time spent at informal and formal meetings and communications
Monitoring and enforcement costs

- *Implementation costs.* Costs associated with implementing and maintaining project interventions or activities including capital expenditures, recurring operations and maintenance expenditures, and staffing costs including project administration and implementation by both landowners and other stakeholders. Example costs include reforestation activities such as the cost of seeds and land preparation, annual maintenance of trees, and staff time or landowner time spent planting and maintaining those trees.

- *Opportunity costs.* Costs associated with foregone benefits or land use due to the initiation of a new activity or project. For example, a landscape project might require that landowners implement a new suite of land interventions, requiring them to forgo migratory labor. Income from migratory labor would then be an opportunity cost. Additionally, changing land use from agriculture to forestry may yield a loss in income and land value from previous agricultural production.

- *Transaction costs.* Costs associated with communications and exchanges. OECD (2013) describes multiple costs that are associated with collective action, defined as "a set of actions taken by a group of farmers, often in conjunction with other people and organizations, acting together in order to tackle local [...] issues." As landscape approaches focus on multi-stakeholder involvement especially amongst resource users, they are rooted in collective action. OECD (2013) notes three types of transaction costs: search, bargaining, and monitoring and enforcement costs. Transaction costs generally relate to the additional time needed to initiate and manage a project or initiative and can be valued using payments for project labor. These costs are frequently included as implementation costs by project funders and implementers. However, labor expenses for project staff and resource users might not be covered by salaries and thus represent an additional expense.

The costs identified above relate to both sectoral and landscape approaches, however integrated landscape management might present unique and/or additional costs. For example, integrated landscape management might have greater transaction costs than sectoral approaches due to increased expenditures to organize local consultations and collection action across relatively large areas. Additionally, integrated landscape management might generate higher implementation and short-term opportunity costs than sectoral approaches. For example, like many sustainable land management interventions, integrated landscape management may require grazing bans or higher labor input to address non-sustainable use of natural resources. This may lead to fewer people migrating for labor and larger areas of land that are taken out of production using exclosures to allow natural regeneration of vegetation or tree-planting efforts that take time to produce results.

Conversely, integrated landscape management can also generate cost savings and increased benefits compared to sectoral approaches in terms of reduced transaction, implementation, and opportunity costs. It is possible, for example, that soil and water conservation measures and sustainable land management activities applied at a watershed or other landscape scale might have high upfront costs but over time, will generate greater economic returns due to restoration of degraded land and increased natural resource productivity, as well as improved ecological connectivity, increased flow of valuable ecosystem services, and greater capacity to properly implement soil and water conservation measures and sustainable land management. Additionally, integrated landscape management might reduce costs associated with conflicts over resource rights due to improved communications. Cost savings are categorized as benefits and are discussed more in depth in the following section.

Benefits Framework

Integrated landscape management for drylands is largely rooted in reversing land degradation and improving ecosystem health and functionality. As such, the benefits of integrated landscape management are intricately tied to the ability of ecosystems within a target area to generate services. Dryland ecosystems are able to provide a variety of economically valuable goods and services. A study by IUCN-ESARO (2010) identified ecosystem services provided by drylands and divided them into four categories: cultural, provisioning, regulating, and supporting services. Examples of these services are described in table 4.2.

These benefits can be categorized as market and non-market benefits. Cultural services are generally non-material but add to well-being, like tourism. Cultural services generate economic value that can be measured based on, for example, transportation costs or entrance costs to national parks. Regulating services are benefits generated by an ecosystem's ability to regulate natural processes such as air and water filtration. Regulating services can be more difficult to quantify, especially if biophysical information on these processes that links them to human welfare is lacking, and often market prices are not available for economic valuation. Provisioning services are benefits that people can directly extract from ecosystems. Generally, provisioning services like food production are easily valued as market prices are available. However, some services like biodiversity and habitat generate non-market benefits. Supporting services are those that underlie provisioning and regulating services and as such are generally not valued in an economic analysis.

Beyond ecosystem services, integrated landscape management also provides social benefits related to investments in social or human capital, health, and improved access to resources and markets. Many landscape management interventions focus on building community institutions, farmer cooperatives, or female-run micro-credit banks. This building of social capital generates multiple

Table 4.2 Ecosystem Services Provided by Drylands in Africa

Cultural	Regulating	Provisioning	Supporting
• Recreation and tourism	• Micro-climate regulation and carbon sequestration • Pollination and seed dispersal • Water and air filtration/purification • Erosion control	• Food and honey • Fodder • Timber and non-timber forest products • Freshwater • Energy • Medicinal and cosmetic products • Habitat	• Soil development • Nutrient cycling • Primary production

Note: Biodiversity in drylands provides the foundation for all four types of ecosystem services. Biodiversity is generally not defined as an ecosystem service.

Integrated Landscape Approaches for Africa's Drylands • http://dx.doi.org/10.1596/978-1-4648-0826-5

market and non-market benefits as it serves to diversify income, improve education and equality, and spread awareness of the value of sustainable land management, which can help to avoid degradation costs in the future.

As noted in the costs section, some social benefits or cost-savings could also be generated due to collective action. OECD (2013) notes that land management at a landscape scale can deliver greater benefits than at the farm scale as collective action can:

- Allow resource users to more easily manage ecosystems across geographical, cultural, and political boundaries.
- Increase uptake of sustainable land-use practices as resource users are more likely to adopt practices if they see their neighbors conducting and benefiting from these practices.
- Make it easier for resource users with different skillsets to collect, share, and create knowledge, skills, and assets at a lower cost.
- Encourage communication and coordination among diverse interest groups and stakeholders, reducing conflict over natural resources, which can result in violence, land degradation, and project disruption. Collective action can improve communications between resource users, reducing the costs of conflict resolution around local issues.

Collective action can result in economies of scale and scope, reducing transaction and implementation costs and enhancing benefits. Mogoi et al. (2009) found that examples of collective action institutions exist across East Africa, including use of traditional indigenous knowledge, conflict resolution, management, and networking. Example benefits of integrated landscape management are summarized in table 4.3.

Table 4.3   Benefits of Integrated Landscape Management

Market benefits

- Improved agricultural, forestry, fuelwood, fodder productivity
- Carbon sequestration
- Avoided transaction costs
- Avoided siltation and flooding costs
- Water quality and quantity regulation
- Pollination services
- Avoided health costs (for example, fewer people migrate to cities, which reduces incidence of HIV/AIDS)

Non-market benefits

- Avoided costs of travel time for water, fuelwood, and other supplies
- Avoided costs of conflict
- Female empowerment
- Increases in biodiversity and improved habitat
- More opportunities for recreation
- Increases in traditional knowledge
- Improved access to health services, markets, and education
- Improved resilience (for example, avoided costs from drought)
- Stronger cultural values

Note: AIDS = acquired immune deficiency syndrome; HIV = human immunodeficiency virus.

Evidence from Landscape Approaches in Ethiopia

This section takes a closer look at landscape approaches operating in Ethiopia, which are described in detail in chapter 5. Over the past 30 years, the Government of Ethiopia's response to land degradation and food security has evolved from a food-for-work focus with top-down administration and mostly technical interventions, to using integrated landscape management principles that emphasize community groups, multi-stakeholder participation, and a focus on watersheds. To illustrate unique costs and benefits from landscape approaches, available literature from drylands restoration programs operating in Ethiopia was reviewed to assess quantitative and qualitative evidence. These programs include the Productive Safety Net Program, Project Ethiopia 2488 and its successor the Managing Environmental Resources to Enable Transition (MERET), as well as the Sustainable Land Management Program. Table 5.2 summarizes the actors, objectives, major activities, and impacts of these programs. The benefits of any individual program are difficult to attribute as programs are implemented in neighboring communities that all contribute to the landscape's ecological integrity. As a result, costs and benefits from a composite of these projects are considered. A more thorough review of the MERET project was possible because it had more data sources.

The MERET project is an interesting case study from a landscape perspective because MERET perhaps best exemplifies the evolution from a food-for-work program targeting Ethiopia's poorest (Project 2488) to a participatory, integrated watershed-based approach (EcoAgriculture 2013a). As of 2009, MERET had covered over 600 sub-watersheds in 74 *woredas*, and had rehabilitated over 40,000 square kilometers of heavily degraded lands (Bewket 2009).

Two systematic evaluations have been conducted of MERET and provide the best evidence for how integrated landscape management principles incorporated into development planning can improve land production and livelihoods both locally and for downstream stakeholders. In 2005, the World Food Programme (WFP) conducted a cost-benefit analysis of the project and in 2012 it commissioned an impact evaluation. Both studies note a lack of baseline data so they relied on qualitative surveys with households, government officials, and WFP members. Both studies also compared sites under MERET with control sites to assess the gain over and above sites with no interventions.

The WFP analysis assessed the benefits of MERET including soil depth, reduced soil loss, soil moisture retention, crop productivity, vegetative cover, woody biomass production, and gully control over a 25- and 50-year time frame. The study included a financial analysis focused on farm-level benefits and an economic analysis focused on economy-wide benefits. Economic costs included investment costs, maintenance costs, and support service costs from WFP. Investment costs for the program consisted largely of the food aid costs, which covered labor input for constructing the various sustainable land management and climate-smart interventions. EcoAgriculture Partners (2013a) estimates that the total cost of the project is almost US$79 million with a 92 percent contribution from WFP and an 8 percent contribution from the Government of Ethiopia.

The WFP (2005) cost-benefit analysis measured the following benefits (see also table 5.2):

- *Reductions in soil loss.* Measured as foregone income or income that could have been earned from cultivated, grazing, and forestland use but was not because of soil loss. The study estimates that the potential production from one ton of soil translates to an annual productivity loss of US$120 per hectare.
- *Soil moisture retention.* Measured as the water-holding capacity of soil. The study found improvements in treated areas.
- *Crop productivity.* This category covers both soil improvements due to increased soil moisture, top soil depth, and fertility, combined with an expansion of cultivated area due to efforts to treat cultivated land to reclaim areas affected by gully and rill erosion and to protect this land from further erosion. The report found that yields were perceived to be considerably higher for treated cultivated lands than non-treated, with an average annual improvement of 1-5 percent, whereas untreated plots were assumed to have a decrease in crop productivity of 1 percent per year.
- *Woody biomass production.* Woody biomass production is essentially the bio-mass produced from afforestation and reforestation efforts. The report found that biomass production increased on treated sites leading to increase of for-est product yields including firewood and pole production.
- *Gully control.* Due to gully control measures, treated areas were made avail-able for cultivation and were also shown to have higher levels of grass growth.

The economic analysis measured both the net present value and the external rate of return. Overall, the analysis found positive results, with an external rate of return of 13.5 percent over 25 years. The report also found an average net present value of US$28,000 for an entire watershed area. The report states: "These figures indicate the project is economically viable even without account-ing for the downstream benefits and intangibles. This is largely due to the mois-ture effect of the soil and water conservation activities on cultivated land." Drier areas were also shown to have a higher external rate of return than moist and humid areas because the effects of water conservation have a greater impact. Finally, forestry activities (for example, natural regeneration in area exclosures, private and community woodlots, and nurseries) were found to have a longer return on investment as it takes time for stands to grow and generate market products.

In 2012, an impact evaluation was commissioned by WFP to better under-stand the impacts of MERET on social well-being, as well as the broader community and ecological impacts (Sutter et al. 2012). Major findings of the report include:

- MERET households have seen an improvement in food security and increased income due largely to improvements in agricultural productivity and support for income diversification. Improvements in agricultural productivity are largely attributed to soil and water conservation activities including terraces, bunds, check dams, and other flood and erosion control interventions which have improved water table levels, soil fertility, and cultivable land area.
- MERET households are better able to manage shocks (for example, drought) due to improved resilience. The report found that MERET households are more resilient as they have multiple income sources, are able to produce more than 1–2 crops, and practice good land management. Additionally, its findings indicate that MERET households have a wider variety of preparation and adaptation strategies than control sites.
- The report found that households' ability to sustain benefits over time was strong.

While not quantified, the WFP (2005) and Sutter et al. (2012) studies, along with a study by Cohen, Rocchigiani, and Garrett (2008), found evidence of off-site or neighboring community adoption of sustainable land management activities and increased economic benefits. Sutter et al. (2012) found that many of the control sites have seen the diffusion effect of MERET interventions. The WFP cost-benefit analysis (2005) stated that they found evidence of numerous downstream benefits generated by MERET, including: reduction in top soil loss caused by wind and rain erosion; reduction in sedimentation of water bodies; an increase in cultivable area; improvements in water table levels; increased soil moisture infiltration; reduced downstream flooding; enhanced carbon sequestration; and soil, flora, and fauna biodiversity benefits. Cohen, Rocchigiani, and Garrett (2008) and Sutter et al. (2012) found evidence of diffusion of MERET benefits, noting that the success of MERET sites led neighboring communities to adopt MERET practices including grazing bans and planting trees.

Other studies have also pointed towards the success of MERET. Nedessa and Wickrema (2010) note that MERET's promotion of a partnership between the WFP and the natural resource extension system has helped to encourage innovation by developing technologies that adapt international conservation engineering standards to the Ethiopian context (for example, sediment storage dams, check dams, and reshaping techniques). Nedessa and Wickrema (2010) state that MERET has improved community empowerment and social capital. Another testament to MERET's success is that several of its principles have been adopted by other programs in Ethiopia, including the Productive Safety Net Program (PSNP) and the Sustainable Land Management Program (SLMP). Evidence from these studies was used to construct table 4.4.

Table 4.4 Integrated Approaches in Ethiopia: Evidence of Costs and Benefits of the MERET Program

Costs	Examples	Quantitative/qualitative evidence for Ethiopia
Capital expenditures on equipment and infrastructure	• Total investment costs to support household labor for technical, social, and environmental interventions	• Investment cost total from Government of Ethiopia and WFP was $79 million[a]; unclear on whether this covers labor costs for administration and implementation.
Annual operations and maintenance (O&M) costs	• O&M costs for interventions to households and funders[a]	• Some O&M compensation paid to farmers for maintenance of existing structures.[b] • Some O&M costs are covered by farmer at no cost and without external assistance.[b] • Physical structures and some biological structures appear to be of superior quality to non-MERET sites including some PSNP sites.[c]
Labor costs for administration and implementation	• Staff time for GoE, WFP, household labor, and other stakeholders • Support service costs	• Investment costs include labor input for constructing SWC and forestry measures[b]: • Soil bund: 150 person/day/km • Stone bund: 250 person/day/km • Fanya juu: 200 person/day/km • Hillside terrace: 250 person/day/km • Bench terrace: 500 person/day/km • Check dam: 0.5 person/day/m^3 • Cut-off drain: 0.7person/day/m^3 • Waterways: 1 person/day/m^3
Foregone income from previous activities and land uses	• Foregone income during physical asset construction • Foregone income from grazing ban • Foregone income from changes in land use • Foregone income from migratory labor which has dropped due to improved local land productivity	• During grazing bans, there is a possible loss of income from livestock activities. • MERET reports an increase in cultivable area from marginal lands which were likely unproductive, thus are not likely to result in opportunity costs for changes in land use.
Transaction costs: Search costs Bargaining costs Monitoring and enforcement costs	• Cost of identifying participants and international and national managers, experts and consultants • Communication expenses (WFP) and time spent at community meetings and discussions • Monitoring and evaluation costs (included in total investment cost)	• Transaction costs unclear, but reports cite potential future costs. • The division of responsibilities for technical support, field supervision, approval of completed work, quality assurance, monitoring, reporting, and trouble-shooting needs to be clarified.[e] • The monitoring system has not been able to provide information on MERET's overall progress due to lack of baseline data, missing indicators or insufficient data collection on certain indicators.[e]

table continues next page

Table 4.4 Integrated Approaches in Ethiopia: Evidence of Costs and Benefits of the MERET Program *(continued)*

Costs	Examples	Quantitative/qualitative evidence for Ethiopia
Benefits		
Improved production/yields from sale of marketable products (for example, crops, biofuel, wood products, fodder)	• Improvement in crop yields • Improvements in woody biomass production and forest product yields including firewood and pole production • Improvements in grass and fodder production • Improvements in productivity outside of project areas	• In 2012, 70 percent of MERET households reported increased income due to improved farm productivity.[f] • Land productivity is valued at $0.02 per ton of soil or $120 per hectare.[b] • Average annual improvement in crop yields between 1 and 5 percent.[b] • Increase in cultivable area from previous marginal lands of 0.2 hectares for MERET households.[b] • Household survey results reported yields were perceived to be considerably higher on treated cultivated lands with soil and water conservation measures, but the extent depended on the general rainfall pattern.[b] • Household income improved, on average, by US$105 per year.[c] • Average standing woody biomass of 20,900 m^3 or 90 m^3/ha. Survival rate of trees was between 65 percent and 85 percent.[b] • The majority of treated sites had a medium level of vegetative cover (50 percent coverage).[b] • Households who diversify production with MERET support in southern Ethiopia will typically inspire another 200 farmers to adopt some of their practices.[g] • Irrigated area increased by 26 percent through 2008.[e]
Benefits		
Avoided negative off-site externalities	• Avoided cost of erosion (for example, cost of siltation in water bodies) • Avoided cost of downstream flooding	• Average improvement of 228 tons of soil/ha/year saved for marginal areas that received MERET support.[b] • Reduction in the accumulation of eroded soil and soil loss that could be washed away from the watershed to downstream areas.[b] • Annual cost of soil erosion to Ethiopia estimated at US$1 billion per year.[d] • Downstream flooding reduced by at least 50 percent.[b]
New income sources and new business development	• Business expansion to beekeeping, plant nurseries, fish production, and poultry production	• MERET has achieved its desirable targets for developing alternative livelihood options including fruit and vegetable production, grasses sales, seedlings from nurseries and woodlots, livestock fattening, and poultry production and sales.[e] • MERET has not yet demonstrated improvements in credit and savings.[c]

table continues next page

Table 4.4 Integrated Approaches in Ethiopia: Evidence of Costs and Benefits of the MERET Program (continued)

Costs	Examples	Quantitative/qualitative evidence for Ethiopia
Carbon sequestration	• Increased carbon sequestration due to afforestation/reforestation and vegetation efforts	• Success in revegetation and reforestation both on-site and off-site can lead to increased carbon sequestration. However, there are no recorded results for improvements in carbon sequestration.
Avoided transaction costs	• Avoided search, bargaining, and monitoring and enforcement costs	• There is good communication and collaboration amongst the national, regional, woreda and WFP staff[e] which would result in reduced transaction costs and costs of conflict over resource access and rights.
Avoided expenditures	• Reduced expenditures on fuel and food • Avoided expenditures on fertilizer due to composting • Avoided erosion hazard	• Farmers responded that erosion hazards had been reduced because gully/waterway channels were more stable.[b] • More than 75 percent of MERET sites reported using composting.[c] MERET households stated composting helped to improve soil fertility.[b]
Avoided costs of travel time for water, fuelwood, etc.	• Reduced travel time to collect drinking water and fuel	• On average, households save 2 hours per day from water collection equivalent to roughly US$8 per year per household.[d] • Household access to water sources increased by 26 percent through 2008.[e]
Biodiversity and habitat improvements	• Improved soil biodiversity and tree diversity	• On average, each enclosure consists of 5 species of trees and there is evidence that wildlife species are re-emerging in area closures.[b]
Health improvements	• Improvements to human and livestock health • Improved household dietary diversity and food security • Reduced healthcare expenditures/avoided deaths due to improved sanitation, food, and water access	• HIV/AIDS prevention training conducted on 56 percent of sites.[e] • About 40 percent decrease in food shortage reported (from 5 to 3 months per year).[d] • MERET households have been found to have more nutritious diets than control site households at statistically significant levels.[c]
Education and skills development	• Improved knowledge of SLM activities and improved self-reliance	• Evidence found that both MERET sites and non-MERET sites improved their knowledge of SLM activities.[g]
Female empowerment	• Increased involvement of women in decision-making processes, income generating activities, and improved wages[d]	• MERET supports women's engagement in planning and management of conservation activities.[g] • Women are the primary household members responsible for collecting water. Households on average save 2 hours per day from water collection equivalent to roughly US$8 per year per household.[d]

Sources: a. Ecoagriculture 2013; b. WFP 2005; c. Sutter et al. 2012; d. Bewket 2009; e. Riley et al. 2009; f. WFP 2013; g. Cohen, Rocchigiani, and Garrett 2008.
Note: Various programs using good practice principles of integrated landscape management have been operating in Ethiopia including Managing Environmental Resources to Enable Transition (MERET), MERET-PLUS (Partnership and Land Users' Solidarity), Productive Safety Net Program (PSNP), and Sustainable Land Management Program (SLMP), with more recent programs building on the successful elements of previous ones. Table 5.2 lists their major objectives, outcomes, and impacts in more detail. The costs and benefits highlighted for MERET in this table have not been compiled in similar detail for these other programs, but are expected to show a comparable pattern, based on the similarities in programming approaches.

Impact studies of the PSNP and SLMP also indicate success. Berhane et al. (2011) found that PSNP improved food security by 1.05 months and that on average, five years' participation in the program improved livestock holdings by 0.38 Tropical Livestock Units. The study also found that food security benefits and livestock holdings improved if households also participated in a complementary food security program like the Other Food Security Program (which evolved into the Household Asset Building Program). Gilligan, Hoddinott, and Taffesse (2008) conducted an impact assessment of PSNP after 18 months and found that the program had on average little impact. This is partially attributed, however, to low transfer levels.

A recent review (World Bank 2014a) of SLMP found that as of January 2014, the project had successfully rehabilitated hillsides, gullies, communal grazing lands, and farmlands on over 190,000 hectares, benefiting 98,000 households. The review found an increase of 26.2 percent of rehabilitation levels on hillsides over the baseline value due to reduced overgrazing. The review also found evidence of increased soil moisture content, reduced soil erosion, reduced crop losses, improved control of flooding and silting, and increased use of small irrigation systems.

In addition to the economic evidence compiled above, other economic valuations or impact assessments were completed for Ethiopia, providing insight as to the success of these initiatives. A study by Haregeweyn et al. (2012) evaluated an integrated watershed management project in Tigray to estimate the impacts of activities on runoff loss and soil loss due to sheet and rill erosion and gully erosion. Overall, the report found that integrated watershed management interventions in the Enabered watershed in Tigray resulted in an increase in vegetation cover and surface roughness. As a result, the watershed has seen a significant decrease in runoff of 27 percent and a reduction in soil loss of 89 percent. The authors note several co-benefits that resulted from integrated watershed management interventions that they were not able to quantify. These co-benefits included micro-climate regulation, carbon sequestration, and habitat for wildlife. Finally, the study also notes that participation by local stakeholders was key to the project's success.

Table 4.4 summarizes the costs and benefits of integrated approaches in Ethiopia, especially the MERET program.

Integrated Landscape Management Benefits and Resilience

Chapter 2 established that household resilience can be analyzed along three dimensions: exposure to shocks, sensitivity to shocks, and coping capacity or the ability to bounce back after a shock. Measuring or valuing resilience in drylands, however, is a difficult task because landscape-level effects in response to certain shocks are difficult to predict and measure. Sukhdev, Wittmer, and Miller (2014) explain that ecosystem transitions and responses to shocks can be non-linear, which makes valuing landscape-level benefits difficult to model. In addition with an increase in the geographic area, the number of ecosystems (or land-cover types and land users) and the social, economic, and ecological interactions among them usually increases, making the overall system more complex and difficult to assess.

Often, programs will develop indicators to demonstrate improvements in household exposure, sensitivity, and coping capacity for shocks, but the methodology behind analyzing and evaluating resilience is still relatively new. Frequently, resilience is measured by the assets and resources available to a given area that can be mobilized to deal with shocks and stresses (Sukhdev, Wittmer, and Miller 2014; Global Risk Forum 2013). These assets and resources are categorized into one of five sources of capital: human, physical, natural, financial, and social. For simplicity, this approach was adopted to determine whether MERET appears to be improving resilience in Ethiopia. Mayunga (2007) provides a conceptual framework for the five sources of capital, as used in figure 4.1. Resilience can be measured by developing an index of resilience based on appropriate indicators. As data are insufficient to construct an index, MERET was evaluated on whether it had improved common metrics for resilience associated with the five sources of capital.

Integrated landscape management inherently helps to develop social capital by using a participatory approach. The extent to which the other four capitals

Figure 4.1 Conceptual Framework for Measuring Community Resilience

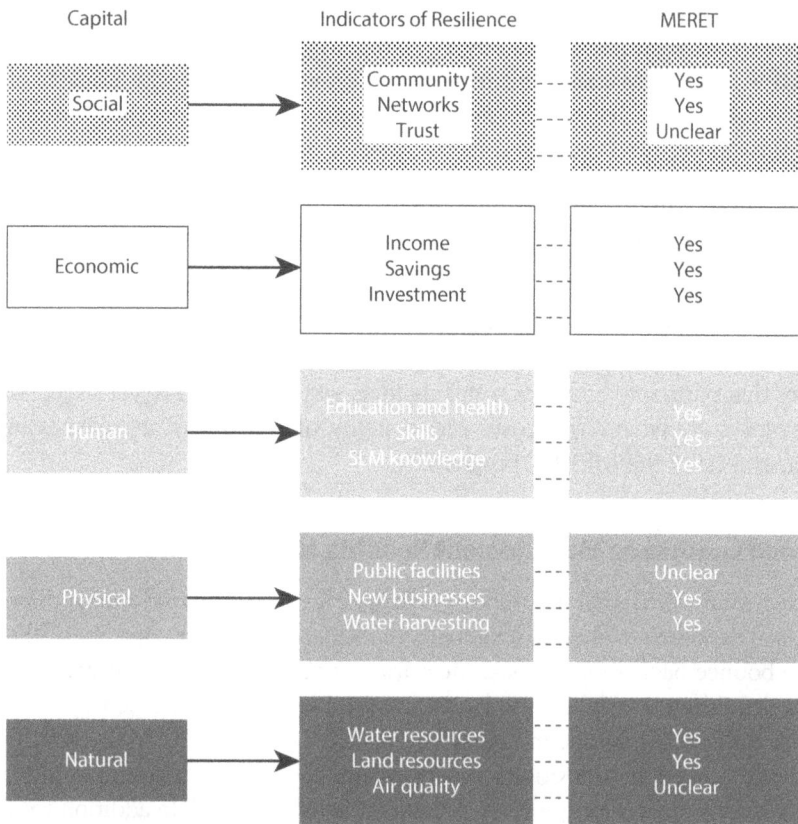

Capital	Indicators of Resilience	MERET
Social	Community Networks Trust	Yes Yes Unclear
Economic	Income Savings Investment	Yes Yes Yes
Human	Education and health Skills SLM knowledge	Yes Yes Yes
Physical	Public facilities New businesses Water harvesting	Unclear Yes Yes
Natural	Water resources Land resources Air quality	Yes Yes Unclear

Note: MERET = Managing Environmental Resources to Enable Transition.

are improved will depend on the individual approach applied and location-specific factors that cannot be generalized.

Based on table 4.4, the benefits of the MERET project are clearly contributing to improvements in each of the five sources of capital for treated watersheds. The 2012 MERET Evaluation Report (Sutter et al. 2012) measured MERET in terms of household and community resilience. Focus groups from the evaluation stated that the most common shocks were drought and flood. The evaluation included an analysis of coping strategies, where MERET and control households were evaluated using a Coping Strategy Index for their ability to respond to these shocks. Overall, the analysis found that MERET households had improved their resilience to shocks compared to control households due to the multiple benefits generated including: improved income diversification; improved income and savings; improved agricultural production; improved water availability and flood control; and improved access to water and sanitation due to soil and water conservation interventions and sustainable land management activities. However, this report also found that while MERET households fared better than control sites in terms of improving food security, they still experienced periods of lacking enough food to meet basic needs. Results of the Sutter et al. (2012) analysis are included in table 4.5. While overall, resilience appears to be improving for MERET households, WFP (2013) and others (EcoAgriculture Partners 2013b)

Table 4.5 Shocks and Preparation Strategies for MERET and Control Households

	MERET	*Control*		*MERET*	*Control*
% of households experiencing any shock in the past two years	*52.0**	*58.4*	*% of households experiencing any shock in the past two years*	*26.1**	*21.1*
N	1,800	1,800	N	1,800	1,800
Type of shock			**Preparation strategy type**		
Drought	34.0	38.0	Household savings fund	44.8	38.1
Crop failure	31.0*	35. 7	Stocked food	37.1*	31.0
Illness of household member	23.3	20.8	Stocked water	1.4	1.1
Flooding	11.8	12.7	Stocked first aid supply	1.1	1.3
Loss of animals	10.0	10.9	Created an emergency plan	5.7	6.1
Death of household member	8.9	6.8	Planted different crops	7.1*	4.0
Conflict	4.1	2.8	Purchased different animals	19.3	18.9
Loss of land	1.4	2.6	Changed livelihood	6.9	8.4
Wind/dust storm	1.1	0.7	Found additional source of income	11.7	12.6
Fire	0.9	0.7	Acquired crop insurance	1.1	.3
Landslide	0.6	0.2	Enrolled in government program	13.0*	21.1
Other	8.7	5.9	Other	5.1*	0.3
	900	983		684	455

Source: Sutter et al. (2012).
Note: MERET = Managing Environmental Resources to Enable Transition.
* $p<0.05$.

cite the need to enhance monitoring and evaluation efforts, and strengthen disaster risk reduction measures.

Implications for Implementing Integrated Landscape Management

Some key lessons learned from current studies across the world and, in particular, Ethiopia include:

- Many countries including India, Ethiopia, Kenya, Niger, and Costa Rica are adopting integrated landscape management principles into their sustainable land management and forest landscape restoration as well as food security and biodiversity conservation strategies. These countries have transitioned from sectoral and top-down, centralized, and bureaucratic approaches to adopting integrated landscape management principles including community participation, improved monitoring and evaluation, and cross-sectoral sustainable land management interventions because they are seeing greater benefits and greater rural household involvement.

- Early Watershed Development projects in India characterized as being technical and using a top-down approach were found to have lower and unequally distributed economic returns compared to more recent projects that employ the latest National Watershed Development Guidelines that promote participatory processes, community capacity building, and other social and environmental interventions.

- Top-down projects in India's drylands provided low returns and were unequally distributed. India case studies of recent Watershed Development projects that include integrated landscape management principles, especially multiple stakeholder involvement and participatory processes, confirmed the cost-effectiveness of integrated landscape management principles and the usefulness of a spatial or watershed approach.

- In Ethiopia, the MERET program has been successful in creating on-farm and downstream benefits by using integrated landscape management principles such as community-based participatory engagement, multiple stakeholders, cross-sectoral collaboration and interventions, and spatial planning.

- MERET has been cited as being more successful than PSNP because it has a stronger focus on promoting community participation and ownership of project interventions than PSNP. As programs in Tigray (for example, PSNP, SLMP, and MERET) contend for funding from the same government and international actors, more people are calling for higher integration of these programs to better streamline and target development funds and build on successful aspects.

- A positive net present value or benefit-cost ratio alone is not sufficient to capture or advance integrated landscape management. Additional studies examining the long-term impacts of integrated landscape management, distributional impacts between primary stakeholders, as well as off-site or downstream costs and benefits are needed to understand how household and ecological resilience is changing, especially in terms of climate change. Other factors that should be considered beyond economic indicators include evidence on opportunities created to change land-user behavior, and uptake and scaling up of integrated landscape approaches and interventions.

- Several of the studied integrated landscape management projects have shown that there is a time lag in terms of achieving a positive economic rate of return and livelihoods improvements due largely to the need to restore ecological conditions such as soil fertility, moisture retention, erosion control, and groundwater levels. As a result, there is a need for financing to support, for example, the establishment costs of improved soil and water management measures, and provide bridge funding until these investments provide positive returns.
- The impacts of integrated landscape management on long-term resilience to climate change are uncertain due to the difficulties in measuring and valuing resilience. However, MERET appears to be improving short-term resilience to climate shocks such as drought.

Despite the need for better quantitative studies for integrated landscape management efforts, existing documentation from economic valuations and impact studies indicates that integrated approaches have the potential to improve resilience in drylands due to improved ecological connectivity, collective action, and improved community engagement and knowledge-sharing.

Note

1. http://www.thegef.org/gef/node/1611.

CHAPTER 5

Case Studies about Integrated Landscape Management in African Drylands

This chapter takes a deeper dive into three case studies that are useful for informing the application of integrated landscape management. Each case study provides some historical context and background summarizing baseline conditions in the study area. The description of each case underscores the importance of a number of key principles of integrated landscape management, which is then followed by a discussion of outcomes and impacts in terms of economic, social, and ecological changes. Finally, each case study concludes with insights and lessons learned. All case studies underscore the importance of learning-by-doing and achieving short-term household-level benefits to drive long-term changes in behavior and drylands productivity. Most importantly, these cases illustrate that it is possible to restore landscapes and increase resilience in the drylands.

The three cases were selected because they help to illustrate the elements outlined in the conceptual definition of integrated landscape management provided in chapter 2, and because they represent different contexts, motivations, and stakeholders. Additionally, the case studies represent a variety of interventions and approaches including soil and water conservation measures, sustainable land management, climate-resilient or climate-smart interventions, and social interventions targeting community capacity building and knowledge. All of them deal in some way with the restoration of productivity and household-level resilience in the drylands of Africa. While each of them has evolved to some extent in its approach over time, none represent a complete example of the application of all 10 principles and good practices of integrated landscape management. It is important to note that other examples of successful landscape-level interventions exist in Africa, but are not fully documented in this book. They include the conservancy program in Namibia, which has made significant progress in scaling up community-based wildlife management (World Resources Institute et al. 2005), the establishment of wildlife corridors through Payments for Ecosystem Services schemes in Kenya (Kristjanson et al. 2002; Gichohi 2003), and South Africa's landscape approach to conserving biodiversity (Cadman et al. 2010). While the case studies described below had perhaps the best evidence base compared to other examples of landscape approaches being employed, they do not cover comprehensively all production systems in the drylands, especially pastoral livelihood systems. A successful application of the principles of good practice for integrated landscape management can boost resilience of pastoral production systems, but a more detailed case study is needed to highlight the specific constraints and opportunities associated with this approach within areas dominated by pastoralism.

The first case study explores the Tigray region of Ethiopia, an area that has seen a shift in government approaches to combating land degradation from food-for-work programs to integrated watershed management and landscape approaches. The second case study examines farmer-managed natural regeneration (FMNR) efforts in Niger, a bottom-up, grassroots approach that reduced cropland degradation through changes in the perception of ownership of on-farm trees, tenure, and agroforestry interventions, enabling landscape-level transformations of "re-greened" agricultural lands. The third case study focuses on improved land and water management efforts in Kenya along the Upper Tana River Basin.

The final section in this chapter provides a side-by-side comparison on how and to what degree the 10 principles of good practices were applied in the three countries. It concludes with insights from the three case studies that are relevant when developing new integrated landscape management initiatives.

Ethiopia Case Study

Context and Background

The case of Tigray is of great interest because it is one of the few regions in Africa where a process of large-scale productivity loss and environmental degradation in the 1970s and 1980s has been reversed over time due to significant investments in landscape-scale restoration. This case study describes Tigray's history of land degradation and the evolution of restoration programs and policies from top-down, centralized approaches to community-based watershed-scale approaches that have resulted in landscape-wide transformation. Restoration efforts in Tigray are also relatively well studied compared to other drylands, providing some of the best ecological and economic evidence.

The State of Tigray lies in northern Ethiopia and is home to over 4.3 million people. Average rainfall in Tigray is between 500 millimeters and 800 millimeters per year, and the region suffers from erratic rainfall and frequent droughts (Gebrewihot and van der Veen 2013). Most soils are characterized as shallow, poor in nutrients, having low organic matter content, and a low water-holding capacity (Gebrewihot and van der Veen 2013). About 53 percent of Tigray's total land area (53,000 square kilometers) is lowland (less than 1,500 meters above sea level), 39 percent is medium highland (1,500–2,300 meters above sea level) and 8 percent is upper highland (2,300–3,000 meters above sea level). Ethiopia's midlands and highlands are characterized as having mixed farming systems and agro-pastoral systems. Pastoral systems are more dominant in the lowlands (Evans, Giordano, and Clayton 2012). About 90 percent of agriculture in Ethiopia is rain-fed as opposed to being irrigated (Nedessa and Wickrema 2010).

Tigray has a history of severe land degradation, especially in the highlands where 50 percent of the area is severely degraded, making it one of Ethiopia's most food insecure regions. Over 80 percent of residents are reliant on agriculture for their livelihoods but about 46 percent of its cropland suffers from severe erosion due to unsustainable management practices (GoE 2012). The average farm size per household in Ethiopia is 0.96 hectares, and 0.91 hectares for Tigray, and is declining, indicating enormous land scarcity, which can lead to increased poverty. This scarcity has been exacerbated in the last few years by the allocation of large tracts of flat land to commercial farmers (Headey et al. 2013).

There are several factors that have contributed to Tigray's wide-scale degradation. First, the historic destruction of vegetation on steep slopes for cultivation, grazing, and household energy supply has resulted in serious erosion and land degradation (Pender, Place, and Ehui 2006). This has also led to the historically high rates of rainwater runoff. A compounding factor is that population growth in Tigray has, for many years, been a driver behind the expansion of farmland on land that is marginal to agriculture. The average population growth rate is 2.5 percent and the average population density is 63 persons per square kilometer. Because vegetation is scarce, many farm families use dung and crop residues as a source of household energy, which means that nutrients are lost and soil fertility is depleted. Additionally, farmers in the region lack access to good infrastructure and technology so they rely on unsustainable and inappropriate land and water management methods (Meikle 2010; Kumasi and Asenso-Okyere 2011; EcoAgriculture Partners 2012b). Political factors have also contributed to Tigray's land degradation. For example, civil war under the Derg regime resulted in additional land degradation as farmers began cutting down trees to make a living and political support for ecosystem restoration was weakened (Kumasi and Asenso-Okyere 2011). Today, the Government of Ethiopia formally owns and regulates the distribution and leasing of land, promotes large commercial farms and resettlement schemes for smallholder farmers, and restricts migration. These policies act as disincentives for smallholders to invest in land (Headey et al. 2013). Finally, a growing threat to the area is that of climate change, which might result in more frequent droughts and more erratic rainfall patterns.

Over the last four decades, the national and regional policy and development response to large-scale degradation has evolved from a top-down, centralized relief approach to a more decentralized, community and watershed-focused participatory approach (Nega et al. 2008; Evans, Giordano, and Clayton 2012). Food- and cash-for-work approaches to address chronic food insecurity and rehabilitate degraded areas through sustainable land management practices (for example, hillside terracing, natural regeneration of vegetation, tree planting, spring protection, small-scale irrigation, on-farm water harvesting infrastructure (ponds), horticulture, and sustainable agricultural techniques) and building of community assets (for example, roads and schools) have been staples throughout Tigray's history, but new programs are aiming to scale up sustainable land management efforts and community asset building efforts and diversify livelihoods. This can be seen through increased integration and coordination of programs, policies, stakeholders, and interventions at the national and regional level.

Efforts to combat degradation and promote conservation in Ethiopia began in the early 1970s with the first Food for Work program beginning in 1971 in Tigray. Efforts were expanded under the Mengistu or Derg regime (1974–91) in response to severe drought and famine (Haregeweyn et al. 2012). Early restoration policies and programs were initiated with support from the Ministry of Agriculture and Rural Development (now the Ministry of Agriculture) and international organizations like WFP and FAO. Under the Derg regime, physical and biological interventions to address degradation included 600,000 square kilometers of soil and stone bunds, 500,000 square kilometers of hillside terraces, reforestation/afforestation including planting of 500 million tree seedlings, banning of tree cutting, and 80,000 hectares of exclosures (World Bank 2014a). Programs in this era focused on providing food relief, generally targeting the most food insecure households, and were considered fragmented.

Project Ethiopia 2488: Rehabilitation of Forest, Grazing and Agricultural Lands (Project 2488), initiated in 1980, is an example of an early project to address land degradation and represented the beginning of large-scale land restoration and soil and water conservation in Ethiopia (Nedessa 2013). The project promoted a top-down, centralized approach for the objectives of food security and building self-sufficiency of the most food insecure area. Activities under Project 2488 focused on food- and cash-for-work, rural road works, and community forestry (WFP 2005). Results of these early interventions were limited because of the quality of interventions, because little attention was paid to integration of activities at the farm-scale and long-term maintenance of interventions, and because there was a lack of ownership and acknowledgment of community rights (Nedessa and Wickrema 2010; World Bank 2014a). For example, some upper ridges were planted with monocultures of eucalyptus trees that drained groundwater and resulted in degradation of soils and water depletion. Additionally, because the Ethiopian economy shrank during this regime due to periods of civil war, agricultural production also shrank due to continued land degradation (Nedessa and Wickrema 2010).

A unique feature of restoration efforts during the 1980s and 1990s in the Tigray region is that they promoted voluntary and free labor (that is, no cash or food provided) out of necessity. The struggle against the Derg required a strong organization of the local population and this proved to be a foundation after 1991 for the development of what is called Mass Mobilization campaigns. Through this unique collective action strategy, every villager in Tigray was expected to voluntarily contribute his/her labor to building public and productive assets. This included, for instance the construction of terraces on steep slopes and the improvement of rural roads, construction of public infrastructure, and irrigation projects (Kumasi and Asenso-Okyere 2011). In a survey of farmers by IFPRI (Kumasi and Asenso-Okyere 2011), respondents stated that their motivations for participating were livelihood improvements, increase in food crop production, and increase in groundwater availability.

After the fall of the Derg Regime in 1991, the regional policy response in Tigray remained focused on food aid and emergency assistance through food- and cash-for-work programs (Gilligan, Hoddinott, and Taffesse 2008), but the Ministry of Agriculture and other stakeholders began promoting bottom-up, participatory, community-based approaches to implementation. By 1992, for example, Project 2488 began using a "local-level participatory planning approach." This approach, developed by the Ministry of Agriculture, FAO, and WFP technical staff, served as national guidance for development agents. The guidelines promoted participatory, community-based watershed development focused mainly on integrated sustainable land management interventions, sustainable intensification of land production, and small-scale community asset building (for example, water ponds). Popular interventions promoted through restoration programs included soil and water conservation infrastructure (for example, stone terraces, soil bunds, microdams), exclosures[1] to restrict grazing, community woodlots, regulations for grazing lands, and application of organic fertilizers (Kumasi and Asenso-Okyere 2011). The local-level participatory planning approach became a major component of land restoration efforts in Ethiopia (Nedessa and Wickrema 2010).

Project 2488 has undergone several expansions since its inception to address early criticisms. For example, a mid-term evaluation of Project 2488 remarked that the program paid little attention to the long-term maintenance of soil and water conservation interventions. Additionally, the evaluation criticized the project's targeting approach, stating:

> The Vulnerability Analysis and Mapping exercise being undertaken in Ethiopia may help to make targeting of the neediest more precise. The project's justifiable target group is so large, however, that it is quite impossible to assist the entire group within the limit of the available resources. In this light, it becomes most important to concentrate in a coherent rather than the currently apparently piecemeal approach of the project's land conservation (e.g., catchment areas), and rehabilitation activities in order to provide a permanent solution to the problems (WFP 1997).

As of 2002, Project 2488 became known as the Managing Environmental Resources to Enable Transitions Program (MERET). MERET still promotes food-for-work schemes to continue building natural and community assets, but it moves beyond food aid to promote an integrated watershed approach using guidance from the local-level participatory planning approach to implement environmental and social interventions such as livelihood diversification (for example, beekeeping and horticulture) and to support the establishment of community management organizations and small-scale irrigation practices, as well as soil fertility management techniques. MERET is perhaps the best example of how restoration strategies have moved beyond sectoral, centralized, and top-down approaches. It also adopts more integrated landscape management principles by moving beyond targeting the most food insecure households to using a more inclusive approach that targets a watershed system, with a specific focus on female empowerment (Nedessa and Wickrema 2010; EcoAgriculture Partners 2012b). MERET also aims to move beyond food assistance to providing technical assistance and financial support through micro-credit loans (Nedessa and Wickrema 2010).

In 2005, the Ethiopian Ministry of Agriculture, with support from the World Bank and the German Agency for Technical Collaboration, released the document "Community Based Participatory Watershed Development: A Guideline," which aims to standardize development guidelines and encourage drylands restoration on a watershed scale and using a participatory, community-based approach (Lakew et al. 2005). This guidance influenced future projects including the flagship Sustainable Land Management Program (SLMP), initiated in 2005. The SLMP is administered by the Ministry of Agriculture and is funded by the World Bank, Finland, the European Union, and Germany (GIZ 2014). The program focuses on capacity development for the Ministry of Agriculture and introduces technologies and interventions to prevent erosion and restore degraded areas including exclosures, terracing, crop rotation systems, improvement of pastureland and permanent green cover. The SLMP builds on lessons learned and principles applied from earlier programs, including MERET. For example, the SLMP fosters the formation of user groups to encourage uptake of sustainable land management (SLM) and promotes a landscape scale as it focuses on watersheds. Finally, the SLMP creates more secure rights for households to invest in their land through rights innovation. The SLMP supported the implementation of the Rural Land Administration and Use Proclamation in 2005 which provides a legal basis for rural land acquisition and use by farmers and pastoralists, as well as the transfer of land-use rights. The program also supported the issuance of second-level certificates in some project areas and provided technical and financial support to finalize first-level certificates. To date, the SLMP has restored 180,000 hectares to productive use and has benefited 194,000 households (GIZ 2014).

Further evidence of a shift towards integrated landscape management is seen through the Government of Ethiopia's Food Security Program (FSP), which was launched in 2005, and has shifted from an *ad hoc* response to food security to a more systematic planned response (Berhane et al. 2011). The FSP manages four programs, including the Productive Safety Net Programme (PSNP) which aims to improve food security for Ethiopia's chronically food insecure households. PSNP promotes food- and cash-for-work and food and cash transfers to targeted households. PSNP also has a Risk Financing Mechanism that acts as an additional source of support in times of drought or other shock. PSNP provides cash for up to five days of work per month, and up to six months per year. Households graduate from the program once they build savings and assets that allow them to meet 12 months of food needs. Households with family members unable to work are eligible to receive unconditional cash or food transfers (World Bank et al. 2013). PSNP was modeled and adapted from MERET (Riley et al. 2009). Work elements include rehabilitating land and water resources, and developing community infrastructure (for example, rural roads, schools, and clinics). Early efforts under the FSP were limited and consequently, in 2010, the Government of Ethiopia enhanced efforts in the PSNP and replaced the previous "Other Food Security Program" with the "Household Asset Building Program" to build households' capacity to generate income and assets (Berhane et al. 2011). An evaluation of PSNP by Berhane et al. (2011) found that the program improved food security in Tigray by 0.75 months, and that this value increased if households received five years of payments up to 1.64 months. Moreover, this value was found to increase if households participated in both PSNP and Household Asset Building Program.

Recent studies (Nedessa and Wickrema 2010; EcoAgriculture Partners 2013b) have criticized PSNP, stating that it is diverting funds away from programs like MERET. Unlike MERET, PSNP does not have a strong emphasis on community natural resource management, and lacks incentives to promote multi-stakeholder involvement of natural resource extension agents and experts. Additionally, food and cash transfers work through household entitlements, whereby food or cash transfers are paid when public works are completed. As a result, Nedessa and Wickrema (2010) argue that entitlements do not act as community-based incentives for soil and water conservation.

How Were Principles of Good Practice for Integrated Landscape Management Addressed?

Today, programs operating in Tigray have moved beyond promoting a small set of sustainable land management interventions and food- and cash-for-work focus to promote a wide range of techniques and interventions to reduce erosion, restore degraded land, promote collective action, build human capital, and diversify livelihoods. The suite of interventions implemented today often includes both physical and biological measures of improved sustainable land management (for example, hillside terracing, natural regeneration of vegetation, tree planting, spring protection), small-scale irrigation, on-farm water harvesting infrastructure (ponds), horticulture, and agricultural techniques. By investing in multiple activities across the agriculture, forestry, and water sectors, most watershed interventions moved beyond a single-sector approach. In addition, programs use a participatory approach and promote community resource user groups, which means that the scale of interventions had to be adapted to a scale that communities can handle. In practice this means that the scale of watersheds managed usually ranges from a few hundred to a few thousand hectares. Table 5.1 describes examples of how

Table 5.1   How Were Principles of Good Practice for Integrated Landscape Management Addressed in Ethiopia's Landscape Restoration?

Principle of good practice	Explanation and examples
Landscape goal(s) and multiple objectives	
Common concern entry point	• Most programs in Tigray including MERET, PSNP, and SLMP focus on food security and natural resource and rural infrastructure building as entry points for funding and interventions. • Local-Level Participatory Planning Approach (LLPPA) and Community-Based Participatory Watershed Development (CBPWD) are integrated into restoration programs in Tigray, allowing entry points like food security and livelihoods development to support sustainable, participatory development (EcoAgriculture Partners 2013a).
Multiple scales	• MERET projects are planned on a watershed scale using a holistic watershed approach and interventions are implemented by farmers. MERET projects also focus on better integrating crop and livestock systems and stakeholders through interventions that define use rights and help to regenerate ecosystem health with, for example, the help of exclosures in certain areas. • MERET also had household/farm-level objectives of improving farm-level productivity and household-level income diversification. • SLMP focuses on the watershed scale, but also promotes farm-level interventions.
Multi-functionality	• National guidelines (that is, the CBPWD Guidelines and LLPPA) recognize the need of integrating the agriculture and livestock sectors. Guidelines promote an "area-based" watershed approach where specific areas along rangelands and transhumance routes will be developed following watershed principles. However, at regional, zonal and *woreda* levels, broad watershed units should be delineated within the specific areas identified for interventions (Lakew et al. 2005). • Programs in Tigray operate across sectors to accommodate crop producers and pastoralists. Related watershed management plans look at multiple types of land use and, even though the accent is usually on a range of physical and biological conservation measures, many plans are cross-sectoral.
Adaptive planning and management	
Adaptive planning and management and continual learning	• Most programs include an evaluation component, but there is scope for strengthening learning and adaptive management among all programs. MERET, for example, did not collect baseline data nor does it have a strong monitoring and evaluation (M&E) component.
Participatory and user-friendly monitoring	• MERET has a focus on results-based management which trains community members in measuring results (Nedessa and Wickrema 2010). • Overall, programs in Tigray are weak on the monitoring component and could better leverage participatory processes for both monitoring and evaluation purposes.
Resilience	• WFP has incorporated disaster risk reduction measures into its programs to improve community and household resilience (EcoAgriculture Partners 2013a). • MERET has integrated livelihoods diversification components into its program to help build resilience (EcoAgriculture Partners 2013a). • Programs show improvements in food security even after dry periods.

table continues next page

Table 5.1 How Were Principles of Good Practice for Integrated Landscape Management Addressed in Ethiopia's Landscape Restoration? *(continued)*

Principle of good practice	Explanation and examples
Collaborative action and comprehensive stakeholder engagement	
Multiple stakeholders	• MERET, SLMP, and PSNP all promote the development of community resource user groups. • MERET employs the LLPPA that targets larger-scale communities as opposed to targeting those most vulnerable within each community. Actions are designed to achieve multiple benefits. • PSNP brings together government, donors, and non-government stakeholders in an integrated manner to build community assets (MoARD 2011). • SLMP is administered by MoA but interventions are implemented using a bottom-up approach involving *woreda* and *kebele* Watershed Development Committees (WDCs) and communities (World Bank 2014). SLMP also adopted mechanisms to encourage participation by marginalized groups including youth, women, and vulnerable groups. • MoA 2005 Community Based Participatory Watershed Development guidelines were created to promote community participation in development decisions.
Negotiated and transparent change logic	• Through MERET, SLMP, and PSNP, stakeholders agree on the need to restore the productivity of degraded land and that village communities play a key role.
Clarification of rights and responsibilities	• The Rural Land Administration and Use Proclamation from 2005 provides rural land users with legal basis for land acquisition, providing an incentive for farmers and pastoralists to invest in their land. The SLMP supported the proclamation in its project sites through registration and certification. In terms of land titling, over 36,626 land parcels were surveyed and 40,227 second-level certificates were issued (World Bank 2014). • Grassroots-democracy decision making and election of accountable leaders are increasingly replaced by fairly standardized regulations (Gebre Michael and Waters-Bayer 2007). Experiments are conducted with allocation of restored communal land to landless youth. This is positive from an equality perspective, but Gebre Michael and Waters-Bayer (2007) remark that the other side of the coin is that privatization of rights to use common land reduces the access of other (poor) farmers to resources.
Strengthened stakeholder capacity	• Capacity building of community members is key for most programs. MERET, for example, provides community management group support to strengthen local voices in land management decisions. MERET also provides technical assistance for natural resource experts. • There is scope for further strengthening of community-based organizations for MERET and other programs. Additionally, a criticism of MERET has been that watershed committees often have to start from scratch, so there is little knowledge-sharing among communities on user-group resource development and planning.

Note: MERET = Managing Environmental Resources to Enable Transitions Program; PSNP = Productive Safety Net Programme; SLMP = Sustainable Land Management Program; MoA = Ministry of Agriculture.

programs in Ethiopia that are operating in Tigray exemplify integrated landscape management principles.

Outcomes and Impacts

Since 1991, soil and water conservation activities have been undertaken on 960,000 hectares and 1.2 million hectares are under exclosures (Rinaudo and Admasu 2010). Several studies have analyzed land use change in the Tigray region (Meire et al. 2012; Kassa 2013). Belay et al. 2014 analyzed land use and land cover changes in Tigray from 1965 to 1994 and 2007. One of their findings is that between 2001 and 2009, areas of central and eastern Tigray, which saw significant investments in land and water management, were greener than what could be expected on the basis of rainfall. Western Tigray, on the contrary, continues to experience degradation. This is most likely due to migration of people from central and eastern Tigray into western Tigray, which has much lower average population densities. The De'sa natural forest in western Tigray is suffering from settlement by migrants and from woodcutting for the production of charcoal.

Meire et al. (2012) used remote sensing data to analyze land use change in northern Ethiopia over the past 140 years and found improvements in woody vegetation and development. The authors conclude that improved land and water management in northern Ethiopia has led to environmental rehabilitation.

A key conclusion of a recent study in this region, which analyzes landscape changes between 1975 and 2006 by comparing photos of a large number of sites, is that:

The recent active intervention by authorities and farmers to conserve the natural resources in Tigray has led to demonstrated significant improvements in terms of soil conservation, infiltration, crop yield, biomass production, groundwater recharge and prevention of flood hazard (Nyssen et al. 2007).

For the programs mentioned above, some specific achievements are provided in table 5.2. Economic benefits of drylands restoration in Tigray are explored in depth in chapter 4.

Insights and Lessons Learned

The Tigray story serves as an interesting integrated landscape management model for restoration of Sub-Saharan African drylands in that its evolution from top-down, centralized programs to more bottom-up, decentralized, and participatory programs shows how a stepwise or sequential series of steps can help garner local and international support and scale up sustainable land management interventions. In Tigray, well-organized communities and a significant contribution of voluntary labor helped to reverse some of the region's early degradation and spread awareness on the importance of ecosystem health. Programs during the last decade were able to learn from early lessons of interventions in the 1990s by adopting a participatory approach to SLM implementation. The national and regional governments, specifically the Ministry of Agriculture, adopted the participatory principle and also moved to decentralize intervention implementation, improve land tenure policies, and implement interventions on more applicable

Table 5.2 Integrated Approaches Operating in Tigray, Ethiopia

Program	Actors	Objectives/Goals	Major Components/Description	Outcomes/Impacts[a]
Managing Environmental Resources to Enable Transition (MERET) and MERET-PLUS (Partnership and Land Users' Solidarity) (2002–present)	• WFP • MoARD • Ethiopia Natural Resource Department extension system • International donors	• Food security • Watershed revitalization • Diversify income • Increase resilience • Female empowerment • Self-reliance • Knowledge-sharing and partnership strengthening • Resilience • Sustainable land management	• Rainwater harvesting • Soil and water conservation measures • Seedling production • Access road construction • Livelihood diversification through beekeeping, horticulture, poultry, dairy, and fish • Revegetation of hillsides	• Increased production of crops and livestock • Nearly 90 percent of beneficiary households reported higher food availability (2008) • Financial rates of return were on average 12 percent for main activities • Time savings for collection of fuelwood, fodder, and water of 2–2.2 hours/day • Improvements in water quantity and quality • Improved soil depth and fertility, and reduced soil erosion • Prevented siltation and flooding to downstream users • MERET households earned 1,226 Birr more than control households over the previous 12 months and nearly two-thirds of all MERET households state that they have escaped from poverty during the last 10 years (compared to less than half of control site households)
Productive Safety Net Program (2005–present)	• WFP and MoA • Ministry of Finance and Economic Development • Disaster Risk Management and Food Security Sector • Save the Children • Other development partners	• Food security and self-reliance through food and/or cash transfers to most food-insecure communities • Create community assets • Encourage household investments • Promote market development	• Food/cash for construction/rehabilitation of productive assets (for example, rehabilitation of land and water resources; community infrastructure including roads, schools, etc.) • Direct support for those not able to work	• Almost 500,000 have graduated from PSNP between 2008 and 2012 • Food security improved by 0.75–1.64 months in Tigray • Improved resilience to climate shocks as people are better able to raise funds in an emergency • Increase in forestry activity and credit access • Significant increases in vegetative cover and broader diversity of plant species • Increase in irrigated area • Reduction of surface runoff and increased infiltration of rainwater

table continues next page

Table 5.2 Integrated Approaches Operating in Tigray, Ethiopia *(continued)*

Program	Actors	Objectives/Goals	Major Components/ Description	Outcomes/Impacts[a]
Sustainable Land Management Program (SLMP)	• GIZ • World Bank • Finland • EU • MoA and regional, *woreda*, and *kebele* level governments	• Reduce land degradation • Improve agricultural productivity of smallholder farmers	• Support to integrated development planning • Strengthen enabling environment (policy and land tenure) • Establish mechanisms to scale up best practices • Land monitoring system • Program coordination and management • Enclosures, reforestation/ afforestation, trenching structures	• Over 190,000 hectares of degraded communal and individual lands treated and managed • Over 194,000 households benefited from SLMP interventions • Over 36,626 land parcels were surveyed and 40,227 second level certificates were issued resulting in an increased level of tenure security among the farmers (more than 98 percent of farmers interviewed) and in the reduction of dispute/conflict (70.5 percent) among land holders
Project 2488: Rehabilitation of Forest, Grazing and Agricultural Lands (1980–2001)	• MoA • WFP • FAO	• Improve food security • Build self-sufficiency of farmers	• Food/cash for construction/ rehabilitation of productive assets	• Creation of local conservation plans (531 plans by 1996 involving 374,000 food-for-work participants) • By 1996, 700 tree nurseries supported

Sources: WFP 2005; WFP 2013; Nedessa 2013; World Bank et al. 2013; Andersson, Mekonnen, and Stage 2009; MoARD 2011; World Bank 2006a; World Bank 2014a; Sutter et al. 2012.

Note: WFP = World Food Programme; MoARD = Ministry of Agriculture and Rural Development; MoA = Ministry of Agriculture; PSNP = Productive Safety Net Programme; SLMP = Sustainable Land Management Program; EU = European Union; FAO = Food and Agriculture Organization of the United Nations; FFW = Food for Work.

a. As many reports do not state benefits specifically for Tigray, these success indicators are for Ethiopia as a whole unless noted.

scales (for example, watershed). MERET in particular has achieved wide success although its funding has been cut in recent years. A cost-benefit analysis of the program conducted in 2005 found that on the farm level, investments in soil and water conservation measures were beneficial in the long-run in terms of improving soil fertility and productivity, but that returns for family labor declined from the establishment year because needed labor contributions were considerable. On the landscape scale, the report also found that benefits included increased groundwater levels and prevention of soil erosion (WFP 2005).

Niger Case Study

Context and Background

The case of Niger illustrates how landscapes can be transformed by a largely bottom-up process driven by innovation and behavior change among smallholder farmers. The spread of the practice of FMNR of trees and shrubs on cropland, the widespread adoption of improved soil and water management practices, and the rebuilding of agroforestry parklands across large areas in the agricultural landscapes of southern Niger are particularly significant in terms of the economic and ecological benefits generated. Much can be learned from an analysis of the landscape-scale transformations of croplands in Niger that have directly contributed to the increased resilience of rural communities. Many of these insights are being applied in efforts to promote FMNR and related sustainable land management practices across agricultural landscapes and to replicate and scale up Niger's success with FMNR and landscape restoration in other countries including Burkina Faso, Ethiopia, Kenya, Malawi, and Senegal. The Niger case study highlights the validity of many principles of good practice associated with integrated landscape management. The experience in Niger suggests that successful scaling of FMNR and improved soil and water management to address key drivers of land degradation such as high rates of rainfall runoff, declining soil organic matter, and loss of soil fertility can serve as an entry point for integrated landscape management aimed at boosting crop yields, improving food security, and increasing household-level resilience.

Niger is among the dryland countries of Sub-Saharan Africa that face major challenges of widespread rural poverty, rapid population growth, political instability, land degradation, and climate change. Niger is one of the lowest ranked countries on the UNDP human development index,[2] and 60 percent of the population survives on less than US$2 per day. Niger has a population of 17 million, with a very high annual growth rate of 3.6 percent. The infant mortality is among the highest in the world, and only 15 percent of women can read and write. Half of the population is under the age of 15 (Pye-Smith 2013).

Niger was badly affected by drought in the late 1960s and early 1970s, again in the mid-1980s, and also in 2005 and 2010. Periodic droughts contribute to a lowering of water tables and shortages of potable water, as well as crop failures, loss of livestock, and increased food insecurity. From the 1950s to the 1990s, deforestation was widespread, driven by population growth and the conversion of savanna woodlands and steppes to cropland. Agricultural development programs and extension services also encouraged farmers to "clean" their fields of trees and shrubs to make it easier to adopt animal traction, plant monocultures of crops in evenly spaced rows, reduce shading and competition for light and water, and adopt other prescribed agricultural practices.

State ownership and control over classified forest reserves and trees on farmlands also contributed to deforestation; without an effective field presence by agents of the national forest service, the policies and laws designed to protect valuable trees had the opposite effect. Local communities harvested what they could without any management rights or incentives for sustained yield management of community forests or trees on farms. As Niger became progressively deforested, the response was to promote large-scale tree planting programs. Some 60 million trees were planted over a 12-year period in the 1980s and 1990s, but the government-led reforestation efforts had a low success rate, with less than 20 percent of the trees surviving. Where trees did survive for a while, there were few effective controls on the cutting and theft of wood. The widespread reduction in tree cover in agricultural landscapes contributed to shortages of fuelwood as well as high rates of rainfall runoff, wind erosion, a decline in the fertility of farmland, and reduced resilience of rural households.

In the mid to late 1990s, the situation had begun to change dramatically, although its full extent would not be recognized for another 5–10 years. By 2005, evidence began to emerge that on some 5 million hectares or nearly half of the cropland in Niger, the density of trees on farms had stopped declining and had instead dramatically increased—at the same time that population density also increased. The land clearing and farming practices of 25–50 percent of Niger's farmers had shifted to adopt the methods of FMNR in combination with other measures to combat wind erosion and reduced soil fertility. Through FMNR, farmers had begun to systematically protect the regenerating shoots of trees and shrubs on farms, and were managing these re-sprouting trees and shrubs to produce a steady flow of wood, fodder, and other non-timber forest products (WRI 2008). By 2010, the increased density of trees on farms was benefiting 4.5 million people, and enabling them to become less vulnerable and more resilient in the face of periodic drought (Cameron 2011). Photo 5.1 and figure 5.1 demonstrate the changes in tree cover between 1955 and 2005 in Niger.

As awareness grew of both the feasibility and benefits of protecting and managing the regeneration of trees in fields, farmers worked to increase the density of trees regenerating from residual root stocks in their fields; trees also grew from seeds stored in the top-soil or re-introduced through livestock droppings. In areas with less sandy soils and high rates of runoff, farmers also recovered degraded land by digging planting pits and adopting other practices to reduce runoff and harvest rainfall. Farmers learned about these improved land and water management practices from innovative farmers in neighboring communities, and also experimented and adapted practices observed during farmer-to-farmer

Photo 5.1 Landscape Dynamics Southwest of Zinder, Niger, 1995–2005

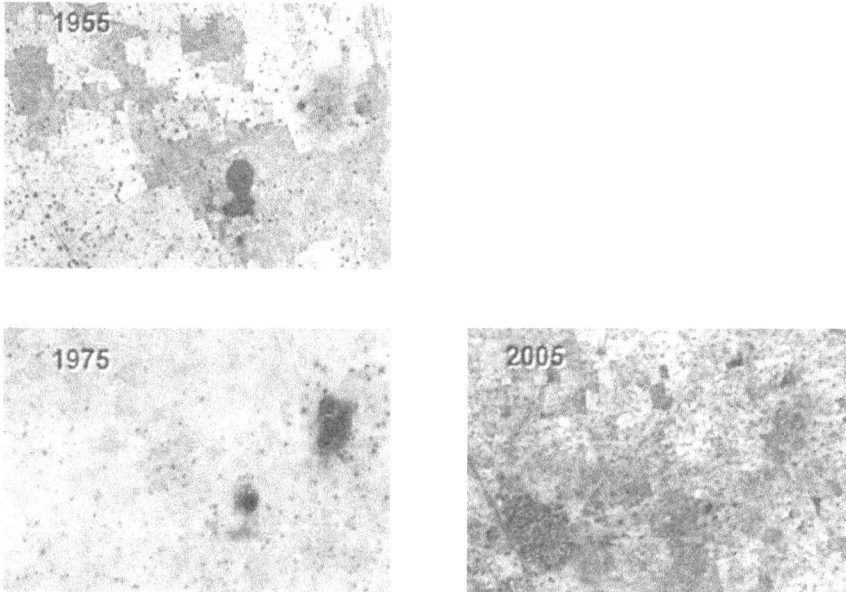

Source: Reij, Tappan, and Smale 2009. Used with permission; further permission required for reuse.

Figure 5.1 Tree Cover Change in Southern Niger, 1955–2005

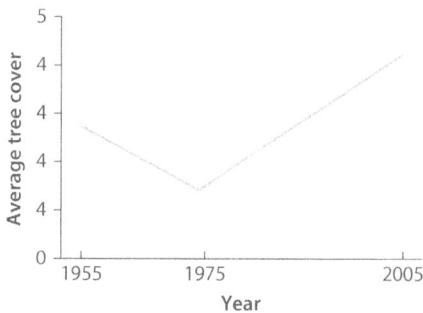

Source: Reij, Tappan, and Smale 2009.

visits and study tours across the region. As a modest investment of time and effort, FMNR contributed to higher crop yields and improved food security, and there was a reduction in migration during the dry season. Farmers also invested more labor, particularly during the dry season, in constructing "half-moons" and in preparing planting pits or improved "tassa" (also known as "zaï" in Burkina Faso).

As the adoption of these improved soil and water management practices was scaled up, farmers achieved greater success both in improving crop yields on existing cropland and in reclaiming degraded and abandoned cropland at the landscape level. Rainwater harvesting practices such as half-moons and *zaï* were very effective in reducing runoff, controlling erosion, and increasing water infiltration, and set the stage for regenerating trees on cropland. As land was reclaimed with rainwater harvesting practices, soil moisture was increased along with soil organic matter. The restoration of soil fertility was accelerated in turn by the regeneration of trees and the combination of water harvesting and agroforestry (see photo 5.2).

In the case of Niger, the transformed and "re-greened" landscape includes large expanses of rain-fed cropland and adjacent land used for irrigated dry season gardens in the lowlands and valley bottoms. Dry-season gardens expanded locally as success with rainwater harvesting contributed to a rise in the shallow water table, and made it more feasible to draw water and irrigate gardens from shallow wells. Land use and management of barren plateaus and brush land used for grazing and harvesting of wood and other products have not been so dramatically affected, although the increased density of trees and production of wood and fodder on farms has reduced pressures to overexploit these lands for firewood collection and livestock grazing.

Photo 5.2   Water Harvesting and Agroforestry

Zaï

Half moons

Source: Chris Reij, WRI. Used with permission; further permission required for reuse.
Note: Farmers who invest in water harvesting techniques also invest in improved soil fertility management. They use manure in the pits and half-moons, and the manure contains seeds from trees and bushes browsed by the livestock. The woody species germinate along with the planted crops. If the farmers decide to protect the young trees that emerge in the same pit as the millet or sorghum, the trees will develop quickly, as they benefit from the combination of harvested rainwater and improved soil fertility. The figure shows a formerly degraded field that was restored to productivity through the combination of water harvesting and agroforestry.

The transformation of agricultural landscapes across southern Niger is particularly notable in the relatively densely populated areas of Maradi, Zinder, and south-central Niger which have relatively sandy soils and are well-endowed with a "seed bank" or root stocks of trees and shrubs that constitute the agroforestry parklands of Niger (see photos 5.3–5.5). In these areas, the woody vegetation on farms is dominated by the economically important gao tree (*Faidherbia albida*), baobab (*Adansonia digitata*), Acacia species, as well as common shrub species such as *Combretum glutinosum*, *Guiera senegalensis*, and *Piliostigma reticulatum*.

Photo 5.3   Extent and Density of Tree Cover Across Southern Niger

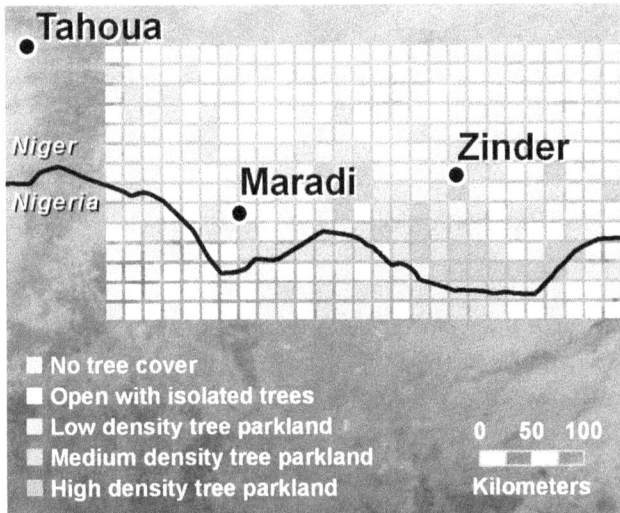

Source: Reij, Tappan, and Smale 2009. Used with permission; further permission required for reuse.

Photo 5.4   Regenerated Gao Trees (*Faidherbia albida*) on Cropland in Niger

Source: Chris Reij, WRI. Used with permission; further permission required for reuse.

Photo 5.5 Restored Agroforestry Parklands in Niger

How Were Principles of Good Practice for Integrated Landscape Management Addressed?

The farmers themselves played a very central role in the "re-greening" and landscape transformations in Niger. A few innovative farmers were generally the "first movers" and they served to demonstrate that the new way of farming with "dirty fields" actually led to greater, not reduced, harvests. Individual farmers were able to make decisions about which species to protect and regenerate and how to manage the increased tree stock for a range of products and services, including improved soil fertility and supply of wood, fodder, and other products for home consumption as well as sale in local markets. No single blueprint or extension package was promoted. This high degree of local ownership and grass roots participation in decision making enabled rural communities to manage trade-offs at both the household and community level.

Community-level collective action was facilitated by the emergence and strengthening of local institutions to govern the use of the expanded tree and shrub resources on farms. Village committees and community-based governing bodies were empowered through decentralization policies to adopt locally enforceable rules that reassured farmers that their investments in protecting and regenerating trees on their farms would not be undercut by uncontrolled harvesting of wood by outsiders. Similarly, farmers were reassured by these local institutions that they had the rights to manage trees on their farms in accordance with their own management objectives, and had ownership over the trees so that they could benefit from the sale of the marketed tree products. Previously, they had to worry about government fines for unauthorized cutting of "protected" species, and payment of permits to transport and sell forest products.

These local organizations, along with supporting NGOs and donor-funded development projects, played a role in supporting peer-to-peer learning and helped accelerate the adoption of FMNR and related land and water management practices through expanded communication and outreach activities.

While a specific landscape-level goal and management objectives were not explicitly established in advance, there was a common understanding that both individual on-farm actions as well as supporting collective actions at village and inter-village level were needed to address the evident problems of deforestation, land degradation, declining soil fertility and reduced crop yields, increased water scarcity, food insecurity, heightened vulnerability to climate change, and other issues. Stakeholder engagement was not driven by a top-down process, but a number of policy and communication initiatives (see section on insights and lessons learned) that helped to strengthen the participation and engagement of rural communities in the widespread adoption of these improved practices, such as FMNR.

Finally, the process of landscape transformation in Niger was aided by adaptive management, as rural communities and supporting organizations developed a step-wise process, beginning with changes in farming practices, and continued to include the strengthening of local institutions to manage the increased tree stocks on farms. In recent years, collaborative efforts have continued to take additional steps in sustainable intensification of agricultural production, such as the introduction of "micro-dosing" and the strengthening of value chains to market firewood from the re-established agroforestry parklands. In the case of micro-dosing, a small amount of mineral fertilizer (for example, as measured in the cap of a soda bottle) is applied directly to the cultivated soil and composted organic matter in the improved planting pit, next to the planted seed. The facilitation of rainfall infiltration and higher soil moisture retention associated with increased soil organic matter content helps to ensure that the fertilizer is taken up efficiently by the crop. And the use of mineral fertilizer complements and enhances the benefits from rainwater harvesting and the restoration of agroforestry systems, contributing to higher crop yields.[3]

The increased capacity of local and community organizations also established a solid foundation that empowered these local actors to plan, negotiate, and adopt rules when a landscape approach was required to establish livestock corridors for pastoralists crossing croplands and FMNR fields in the western Maradi region (Dakoro and Guidan Roumji Arrondissements). Three demarcated livestock corridors have protected farmers' crops and trees, safeguarded grazing and water access areas for herders, and resolved conflicts through agreed upon dispute resolution mechanisms (Byrne et al. 2011; Learning Initiative 2012). Although Niger's Rural Code (1993) and revised Pastoral Code (2010) protect livestock corridors, funding and technical support at landscape scale overcame lack of government resources and facilitated collaboration among farmers, pastoralists, local and regional governments, and NGOs to reduce the incidence and intensity of conflicts and assist livestock movements to markets.

In contrast to the positive results from the largely bottom-up development of "re-greening" initiatives in Niger, the long-term results from an ambitious program to restore large areas of degraded land in the region of Keita are tangible,

Integrated Landscape Approaches for Africa's Drylands • http://dx.doi.org/10.1596/978-1-4648-0826-5

but have not proved to be self-scaling. Considerable technical support was provided along with millions of dollars in food aid to local communities to build water harvesting structures on barren plateaus and slopes, and small dams in valleys. Besides this, the project systematically planted trees along or in all water-harvesting structures. The massive investments "worked" for a time from a technical standpoint, and were successful in reducing runoff and erosion and in restoring the vegetative cover. Following the suspension of project assistance, however, local communities have not continued to invest in the extension of the landscape level restoration and lack of maintenance is a growing problem.[4]

Table 5.3 reviews specific examples from the Niger case study in relation to the identified principles of good practice for integrated landscape management.

Table 5.3 How were Principles of Good Practice for Integrated Landscape Management Addressed in the Niger Case Study?

Principle of good practice	Explanation and examples
Landscape goal(s) and multiple objectives	
Common concern entry point	• Bottom-up participation in the adoption of improved land and water management practices leading to landscape-scale transformations was driven in large measure by a shared concern about declining soil fertility and crop yields, increased vulnerability to drought and erratic rainfall, as well as shortages of firewood and fodder.
	• No specific landscape-level goal and management objectives were explicitly established in advance for FMNR or rainwater harvesting efforts. However, there were cases where support from NGOs such as SIM provided training to farmers and other NGOs like CARE helped farmers to align some of their individual practices in their fields across a larger geographic area (for example, windbreaks in the Majjia valley) and assisted community-level/local institutions in adopting locally enforceable rules to safeguard farmers' investment in trees.
	• More recently, with assistance from the Swiss Development Cooperation (SDC), three demarcated livestock corridors in the western Maradi region (Dakoro and Guidan Roumji Arrondissements) established a landscape-level goal to reduce the incidence and intensity of conflicts between crop farmers and herders and assist livestock movements to markets.
Multiple scales	• Significant farm-level and landscape-level benefits have been achieved. Individual farmers worked at the scale of their farming system or managed agro-ecosystem; rural communities collaborated to work at the village and inter-village scale. Supporting organizations such as IFAD's PPILDA project and SIM worked through networks to facilitate scaling up by investing in farmer-to-farmer visits, extension and training, communication and outreach to raise awareness of the improved practices and associated benefits.
Multi-functionality	• The improved practices contributed to higher crop yields, increased production of firewood, fodder and other products, income generation, and poverty reduction. Across the rural landscape, it resulted in decentralized resource management and strengthening of community-based organizations governing land and resource use, and restoration of a range of ecosystem services.
	• Farmers integrated forest, crop, soil, and water management practices to benefit from the synergies among these practices and aligned efforts in their high-potential fields and the adjacent more degraded lands.

table continues next page

Table 5.3 How were Principles of Good Practice for Integrated Landscape Management Addressed in the Niger Case Study? *(continued)*

Principle of good practice	Explanation and examples
Adaptive planning and management	
Adaptive planning and management and continual learning	• Farmers and rural communities proceed in a step-wise process, starting with changes in farming practices and then building and strengthening local institutions to manage and safeguard the increased tree stocks on farms. There is a need for further strengthening of learning and adaptive management. While technical staff of government agencies are becoming familiar with many aspects of farmer innovations and the cost-effectiveness of the improved practices, more needs to be done to ensure that what has already been achieved in terms of landscape-level transformation is fully accounted for in ongoing sectoral program planning in agriculture and forestry.
Participatory and user-friendly monitoring	• Apart from occasional studies, there are few examples of participatory monitoring and evaluation tools being used.
Resilience	• There is a growing body of evidence that FMNR and related practices are contributing to increased resilience, but additional research would be helpful.
Collaborative action and comprehensive stakeholder engagement	
Multiple stakeholders	• Smallholder farmers are the key stakeholders. Village-based organizations played a key role in facilitating peer-to-peer learning through farmer-to-farmer visits, and in deliberating, adopting and enforcing local rules.
Negotiated and transparent change logic	• A majority of stakeholders agreed on the need to restore the productivity of degraded land. Once trees and shrubs were regenerated, communities established locally enforceable rules for further protection and growth of the trees on farms.
Clarification of rights and responsibilities	• Government policies related to decentralization, security of land tenure, and improved natural resource management were shifting and becoming more supportive of the devolution of the ownership of trees on farms and of management rights to farmers. This shift was reinforced by Forestry Code reforms adopted in 2004.
Strengthened stakeholder capacity	• Assistance with farmer-to-farmer visits, provision of short-term training, and facilitation of peer-to-peer learning all helped to build capacity of key local stakeholders.

Note: FMNR = farmer-managed natural regeneration; NGOs = nongovernmental organizations; IFAD = International Fund for Agriculture Development.

Outcomes and Impacts

As a largely bottom-up process, the landscape transformation in Niger was accomplished, to a great extent, by word-of-mouth exchanges of experience among farmers. It did not require heavy investment in infrastructure or large volumes of externally provided inputs. Trees and shrubs on farms were mainly local species that were protected by farmers and regenerated from rootstocks or seeds in the top soil. Protection and management of these regenerating tree stocks was done by farmers who were working in their fields and it required relative little labor. In the densely settled and semi-arid croplands of Niger, the control of bush fires was not an issue and uncontrolled burning was not a threat to the regeneration of trees and shrubs. However, farmers did need to rely on strong community-based institutions to uphold local agreements on the movement of livestock to reduce damage to seedlings from livestock browsing as well as to prevent illegal cutting of trees by outsiders.

In comparison with the costs of tree planting, which usually shows low survival rates and costs in the order of US$200 per hectare or more, the estimated costs of outreach and facilitation to encourage FMNR amounted to US$10–20 per hectare. The survival rate, as well as the rate of growth of trees regenerated through FMNR, is also much higher than in the case of planted trees.

Benefits from the widespread adoption of rainwater harvesting techniques and agroforestry (through FMNR) accrued at both the household or farm level and the landscape level. Households benefited from improved soil fertility, increased crop yields, higher incomes, and the production of multiple streams of goods and services (see table 5.4).

These improved practices enabled farmers to directly confront some of the key constraints on crop production, including declining soil organic matter content, low soil fertility, and increased soil moisture stress. In the Aguie Department of Niger, farmers who had been harvesting only 150 kilograms per hectare of millet were able to harvest about 500 kilograms per hectare following the adoption of FMNR. And with the application of small doses of fertilizer (micro-dosing), yields have risen to 1,000 kilograms per hectare (Pye-Smith 2013).

In the Illela district and other rural areas of Niger, rainfall varies significantly from year to year and is associated with fluctuations in crop yields. For example, during the period 1991–96, average annual rainfall ranged from a low of 233 millimeters in 1993 to a high of 581 millimeters in 1994. During these periods, average crop yields in the district ranged from 241 to 386 kilograms per hectare (see table 5.5). Of particular interest is the impact on crop yields

Table 5.4   Benefits of Trees on the Farm and Landscape Levels

	Farm-level benefit	Landscape-level benefit
Increased crop yields	✓	✓
Fruit production	✓	
Fodder for animals	✓	✓
Traditional medicine	✓	
Firewood production	✓	✓
Increased soil fertility	✓	
Increased water infiltration	✓	✓
Decreased soil erosion	✓	✓
Reduced wind speed	✓	✓ [possible - threshold]
Decrease in temperature	✓	✓ [possible - threshold]
Habitat for millions of migrating birds	Some birds can be beneficial to farmers for pest control and seed dispersal	✓
Increase in biomass and carbon/contribution to mitigate climate change		✓

Integrated Landscape Approaches for Africa's Drylands • http://dx.doi.org/10.1596/978-1-4648-0826-5

Table 5.5 Rainfall, Water-Harvesting Techniques and Cereal Yields in Niger (1991–96)

Rainfall	1991	1992	1993	1994	1995	1996	Average 1991–96
Badaguichiri	726 mm	423 mm	369 mm	613 mm	415 mm	439 mm	
Illéla	581 mm	440 mm	233 mm	581 mm	404 mm	440 mm	
Zaï							
T0	—	125	144	296	50	11	125
T1	520	297	393	969	347	553	513
T2	764	494	659	1486	534	653	765
Half-moons							
T0	—	86	77	206	28	164	112
T1	655	293	416	912	424	511	535
T2	1183	538	641	1531	615	632	857
Average							
Illéla district	386	241	270	362	267	282ı	301

Source: Hassane et al. 2000.

Note: T0 = adjacent fields; T1 water-harvesting technique + manure; T2 water-harvesting technique + manure + urea

from treatments that combine water-harvesting techniques such as tassa (improved planting pits) and half-moons, along with manure and urea fertilizer. In trials during this same period on relatively infertile fields with average yields of only 112–125 kilograms per hectare, average crop yields during the six-year period rose to a level of 765–857 kilograms per hectare.

In addition to beneficial impacts on crop production from water harvesting and soil and water conservation measures, farmers adopting FMNR also benefit from the production and sale of other products, such as wood and fodder. ICRAF researchers recently estimated that the annual value of fuelwood harvested by farmers practicing FMNR amounted to US$224–256 per household. Researchers have estimated that farmers regenerating 40 trees per hectare can earn an additional US$140 per year—about half the annual income of most smallholder farm households in Niger. Another study found that the wood harvested and sold over a 12-year period from 1985–97 from land where farmers in 100 villages around Maradi had adopted FMNR was worth US$600,000 (Cameron 2011).

Agroforestry systems supply more than wood. Trees produce valuable medicinal products, nuts (which can be consumed directly or processed into edible and non-edible oils), and edible fruits and leaves. The value of leaves harvested from one mature baobab tree amounts to US$34–70 per tree, which was sufficient to buy 70–175 kilograms of grain in 2012 (Reij 2013 using data from Yamba and Sambo 2012). Some farmers have fields with dozens or more baobab trees. A farmer with 2 hectares and 50 baobab per hectare could bring in more than US$2,000 per year in added income (Cameron 2011). ICRAF recently estimated that the value of tree products among households practicing FMNR amounted to around US$1,000 per household per year (Pye-Smith 2013). Other studies have also confirmed an increase in the gross value of crop production for farms practicing FMNR, taking account of cereal production along with cowpeas and groundnuts. Gross income per capita was 86,104 CFA (US$167) for adopters versus 62,996 CFA (US$122) for non-adopters (Haglund et al. 2010, cited in Pye-Smith 2013).

A survey of annual household income for rural households in five villages where new agroforestry parklands had been established through FMNR revealed that incomes had increased by US$37 to US$200 per household—a significant amount for families with incomes as low as US$1–2 per day (see table 5.6). Moreover, households classified as particularly poor and extremely vulnerable were able to boost their incomes from US$45 to US$116 or nearly as much as the least vulnerable households. This indicates how the adoption of FMNR and restoration of agroforestry parklands can be a very equitable approach to increase rural incomes and resilience.

Experience from Niger confirms that FMNR and other improved land and water management practices helped farmers to decrease their sensitivity to drought and reduced the effect of erratic rainfall as they could rely on the harvesting and sale of poles, firewood, fodder, and other products during periods of reduced yields of annual crops. As agroforestry parklands were restored, rural communities benefited from the general effects of sustainable intensification and diversification of crop-production systems.

Initial estimates of increases in crop yields associated with FMNR were in the order of 100 kilograms per hectare, or about 500,000 additional tons of cereal over the 5 million hectare landscape. More recent studies have documented an increase in crop yields of 240 kilograms per hectare when millet is cultivated under a canopy of *Faidherbia albida* trees (Pye-Smith 2013).

At the landscape level, aggregate effects of increased tree density on farms included reduced wind erosion and improved nutrition. As the practice of FMNR spread across millions of hectares, households in southern Niger reported reduced wind speed at the beginning of the growing season, lowering households' exposure to negative wind effects, which can destroy emerging crop seedlings and require farmers to replant multiple times (Larwanou, Abdoulaye, and Reij 2006). Recent modeling studies also suggest that tree-covered landscapes, which have higher evapotranspiration rates, can have a positive effect on rainfall in landscapes

Table 5.6 Average Annual Household Income from Agroforestry Parklands (US$)

Village degree of vulnerability	Kouka Samou	Doukoum Doukoum	Kirou Haussa	Zedrawa	Daré
Least vulnerable	200	40	140	125	135
Medium vulnerable	110	37	120	70	63
Very vulnerable	80	83	26	40	100
Extremely vulnerable	104	50	116	80	45

Source: Yamba and Sambo 2012.

downwind from these tree-covered landscapes (van Noordwijk et al. 2015). The evidence for location-specific cause and effects in Niger, however, has not been compiled yet (Reij, personal communication 2014). Finally, landscape-level re-greening contributed to climate change mitigation with the sequestration of an estimated 30–60 million tons of carbon over 5 million hectares in the past two to three decades (Place and Garrity 2014; Stevens et al. 2014).

Insights and Lessons Learned

No single policy action, project intervention, or investment was sufficient by itself to trigger the re-greening of Niger. Rather, success was achieved through a combination of efforts that collectively contributed to a positive enabling environment (Sendzimir, Reij, and Magnuszewski 2011). Specific barriers that were overcome included:

- *Facilitation of innovation and demonstration of the feasibility of FMNR and zaï* as an alternative to traditional practices of preparing cropland for cultivation of annual crops; staff of NGOs like SIM, but also of a major IFAD-funded project working at the community level, gained the confidence of farmers and worked with them to encourage the protection and regeneration of trees in fields and trials with half-moons and *zaï*, which led to changes in traditional behaviors by revealing the benefits that followed the adoption of these unconventional, improved land and water management practices.
- *Changes in the role of the national government and technical services* from central control and a regulatory framework that discouraged local initiatives, towards increased support for decentralized natural resources management.
- *Change in policy and institutional focus* from protection by the state of remaining trees and woodlands, to enabling regeneration and harvesting of locally managed tree stocks to directly benefit local communities investing in protection and management of trees on farms.
- *Reform of the Forestry Code* in 2004 to officially sanction local ownership and management rights for trees on farms, following what had developed in practice through the adoption of FMNR in the preceding decade.
- *Enabling of market access through investment in roads and rural infrastructure* to facilitate marketing of surplus production from restored and diversified rural production systems (including rain-fed crops, trees on farms, and irrigated vegetable gardens).

Despite the widespread adoption of FMNR and other innovative practices and the significant level of benefits associated with restored agroforestry parklands, many sectors remain focused on conventional approaches to achieving sector targets and continue to overlook the fundamental need to restore and improve the management of land and water as a necessary complement to other interventions such as the provision of improved seeds and fertilizers, and increased investments in wells, pumps, and water storage facilities. These conventional and relatively narrowly focused interventions aimed at boosting crop production or increasing water supply can be made more sustainable and effective when designed with the full participation of local stakeholders and supported as elements of an integrated landscape approach.

Kenya Case Study

Context and Background

This case study presents a multi-sector program to reduce poverty, increase food security, and empower local communities to manage water, agricultural lands, and natural resources sustainably within the Mount Kenya ecosystem. The program combined ecological and socio-economic criteria to geographically target its activities and sought to achieve household, community, and landscape-level benefits. This case study was selected because it applied many of the principles of good practice for integrated landscape management. In addition, it highlights the connection between farm-level investments in improved soil and water conservation, and downstream impacts at watershed scale.

In many cases, investment decisions for improved soil and water conservation are based solely on the costs and benefits calculated at farm level. Structural measures (for example, bench and *fanya juu* terraces, cut-off drains, stone lines) and agronomic and vegetative measures (for example, grass strips, contour farming) need to be profitable for the farmer, and in many locations these measures are profitable after one or two years (Liniger et al. 2011). Since there is a time lag between the initial establishment of these measures and farm-level benefits, smallholder farmers often need support (for example, soft loans) to cover the upfront costs and can benefit from additional farm management advice that results in higher-value crops and market access (Onduru and Muchena 2011). The latter improves the price ratio of farm outputs to inputs (for example, high-value fodder crops planted to stabilize soil and water conservation structures will pay off quicker if fed to improved dairy breed cattle with access to a well-functioning milk market).

If improved water and soil conservation measures are practiced by a large number of farmers, they can have positive downstream benefits such as reduced sedimentation rates, increased water quantity, and improved dry or wet season flows. In the Mount Kenya case study, a landscape perspective that links downstream beneficiaries with upland farmers creates the opportunity to establish a financial mechanism and institutional setup (Green Water Credits) to reward thousands of upstream farmers to invest in and maintain improved soil and water management practices.

Finally, the case study demonstrates the linkages between high-potential croplands and drylands. In the case of the Tana River, dryland irrigators are downstream and depend on the high-potential uplands for a steady water supply. In fact, community-based smallholder schemes in Machakos County are at the tail-end of supplies and are competing with large commercial irrigators, public irrigation schemes, and other water users upstream. Upstream land and water management practices that result in better regulated river flows would supply irrigation water for longer periods, allowing downstream dryland farmers to enhance production or put more land under irrigation.

The upper Tana River basin extends over 17,420 square kilometers and is the home of 4.5 million people. Ecosystems range from humid forests to arid rangelands and include the following major land cover or land uses: national parks and forest reserves extending over a large portion of the two main mountain ranges (Mount Kenya and the Aberdare Range), covering about 20 percent of the region (WRI et al. 2007); high-potential and high-rainfall agricultural lands in the foothills below these mountains including small-scale subsistence and cash crop farming (mostly rain-fed), plus commercial tea, coffee, and horticulture crop production; and the lowlands (about 30 percent of the region) mostly falling within dry sub-humid and semi-arid agro-ecological zone with irrigated crop production (for example, rice, pineapples, vegetables), marginal rain-fed cropping, and livelihoods depending on livestock and rangeland production systems.

The average population density is 300 people per square kilometer. The upper Tana includes some of the most densely populated rural areas of Kenya (more than 1,000 persons per square kilometer) while areas with very low population densities are found in the lowlands, typical of Kenya's arid and semi-arid lands. (It also contains large protected areas with no official settlements.)

Spatial patterns of poverty in the upper Tana reflect very much the gradient of rainfall and agricultural endowment. Administrative units at higher elevations in general have lower poverty rates than those further downstream. The communities in the lower plains and the drier parts of the upper Tana have the highest poverty rates.

The Tana River and its tributaries from Mount Kenya and the Aberdare Range provide the only permanent source of surface water in the southeastern drylands of Kenya's lowlands. The Kenya Electricity Generating Company Limited operates five hydropower stations and is the largest water user. The Nairobi Water Company is the second largest water user (about 75 percent of Nairobi's water comes from the upper Tana River; municipal water demand is projected to grow by 6 percent per year (van Steenbergen, Tuinhof, and Knoop 2011)). Downstream irrigators are the third largest water user. The irrigation schemes total 6,870 square kilometers including Del Monte pineapple, Kakuzi plantations (trees and other crops), Kenya's largest rice production scheme, and smallholder dryland irrigation in Machakos County through the Yatta canal. The Tana River is also an important surface water source in the drylands further downstream, and environmental flows are required to preserve the ecology of the river delta and associated coastal ecosystems.

Major household and landscape-level challenges include demographic pressure with associated small farm holdings and concentrated pockets of high poverty. Low agricultural productivity and market failure of traditional cash crops, and degradation of forests and other natural resources are other pressures that threaten the integrity of the greater Mount Kenya ecosystem (GEF 2004).

This case is based on a literature review of the completed Mount Kenya East Pilot Program (MKEPP) and the new Upper Tana Catchment Natural Resources Management Project (UTaNRMP), which is scaling the lessons learned from MKEPP to the remaining sub-catchments of the upper Tana River basin. MKEPP was piloted for five sub-catchments (completed in 2013/14), and UTaNRMP is covering all 24 sub-catchments of the upper Tana River basin (see table 5.7 for more details on the programs). The literature review also examined background studies on the cost and benefits associated with establishing a Green Water Credit scheme for the upper Tana building on these experiences.

How Were Principles of Good Practice for Integrated Landscape Management Addressed?

Table 5.8 details how the Kenya case study applies major components and associated principles of good practice for integrated landscape management.

Outcomes and Impacts

Benefits at Household and Community Levels

The program sought to improve the livelihood of small-scale farmers who rely on rain-fed agriculture. Considerable investments were made and program outputs delivered in improving water infrastructure (for example, 1,200 hectares of irrigation schemes were designed, constructed and being used and 146 boreholes rehabilitated) and boosting agricultural practices and livelihoods (for example, 16,483 farms established soil and water conservation structures and 137 farmers' field schools were set up).

Agriculture-related income increased after program intervention: 16 percent to 22 percent from agriculture-related employment; 32 percent to 38 percent from small agribusinesses; and 29 percent to 51 percent from horticulture (Capital Strategies Kenya Limited 2012). Most farmers who had adopted improved soil water practices reported an average 65 percent increase in agricultural productivity. About 75 percent of [targeted] households participated in effective water resource management and use, and overall access to water for domestic and productive purposes increased in the pilot area.

Community organizations were strengthened by the program's support to establish and build the capacity of Water Resource Users' Associations (WRUAs) (for example, 17 WRUAs plus five overarching river WRUAs). This in turn advanced WRUAs' capacity to develop sub-catchment management plans, carry out abstraction surveys, and control surface water abstractions.

Ecosystem-Wide and Landscape-Level Benefits

The program sought to achieve a number of ecosystem-wide and landscape-level benefits such as improved water quantity and quality downstream of the intervention areas and restoration of forest cover and degraded forest lands. Not all of these landscape-level impacts were monitored. Some investments in sustainable soil and water management require long-term maintenance and effort at scale (larger than the focal development areas chosen in the five sub-catchments),

Table 5.7 Upper Tana River Basin Case Study Details

Program/Project	Target area and target population	Goal and major components	Objectives	Examples of landscape actions
Mount Kenya East Pilot Project (MKEPP) 2002–14	IFAD (loan) • 5 sub-basins (focus on farmland along rivers) • 580,000 people • 136,000 households (poor and at risk being poor) Total cost: US$25.9 million IFAD: US$16.7 million GEF: US$4.9 million GoK: US$1.8 million Community: US$2.5 million	*Community Empowerment* • Rural communities empowered for sustainable management of natural resources *Sustainable Rural Livelihoods* • Natural resource-based rural livelihoods sustainably improved *Sustainable Water and Natural Resource Management* • Land, water and forest resources sustainably managed for the benefit of the local people and the wider community	• Reduce poverty • Improve food security • Increase incomes • Effectively use natural resources • Improve access and management practices for water resources • Use better farming practices	*Sustainable Water Resources Management* • Formation and strengthening of WRUAs • Rehabilitation of boreholes • Construction of water harvesting structures • Construction of water supplies • Construction of shallow wells • Construction of springs • Improved irrigation systems *Sustainable Natural Resources Management* • Seedling planting • School greening program • Roadside conservation • Community training in tree nursery, participatory forest management, environmental governance, seed collection and handling • Wildlife control fences • Forest rehabilitation • Riverside tree planting • Environmental conservation awards program • Energy saving devices *Rural Livelihoods* • Farmer field schools • Soil and water conservation measures • Water harvesting measures • Improved planting materials • Dairy cow upgrading • Goat upgrading • Poultry upgrading • Rehabilitation of rural access roads • Creation of new roads *Community Empowerment* • Formation of WRUAs, CFAs, FDA CIGs • Female empowerment • Gender sensitization • Training in operation and management of physical assets
Upper Tana Catchment Natural Resources Management Project (UTaNRMP) 2012/13–19/20 Total cost: US$68.8 million IFAD loan: US$33.0 million Government of Spain loan: €12.8 million Communities: US$7.5 million GoK: US$11.0 million	• Mount Kenya ecosystem (National Park [700 km^2], Forest Reserve [2,000 km^2], and agricultural landscape within 10 km of forest reserve) • 17,420 km^2 • 24 sub-basins (12 priority sub-basins and forest areas, 5 MKEPP sub-basins and other sub-basins) • Focal Development Areas (FDAs) ecosystem approach (not administrative units); 5 km stretches on each side of "critical" rivers; environmental hot spots • Target 200,000 poor rural households			

Sources: Africa Livelihoods Networks Limited 2013; Capital Strategies Kenya Limited 2012; Gikonyo and Kiura 2012; GEF 2004; IFAD 2012a; IFAD 2012b.

113

Table 5.8 How Were Principles of Good Practice for Integrated Landscape Management Addressed in the Kenya Case Study?

Principle of good practice	Explanation and examples
Landscape goal(s) and multiple objectives	
Common concern entry point	• To develop a shared perception of the upper Tana as a management unit, the program was formulated to address the key pressures threatening further degradation of the Mount Kenya ecosystem: poverty, pressures on forest reserves, national parks and water resources, and unsustainable farming practices in high-potential agricultural lands. The program defined achievable goals that would benefit the targeted 136,000 households (either poor or at risk of being poor) and provide public goods (improve ecosystem service delivery) in the Mount Kenya ecosystem. The goals were to increase income (reduce poverty), manage water, agricultural lands, and key supply areas of important ecosystem services (for example, hilltops, riparian buffer zones, forest and forest reserves, and national park/mountain zone) more sustainably. The International Fund for Agriculture Development (IFAD) and its partners aimed at achieving the following three major outcomes: – Rural communities empowered for sustainable management of natural resources – Natural resource-based rural livelihoods sustainably improved – Land, water and forest resources sustainably managed for the benefit of the local people and the wider community • The program had a deliberate geographic targeting approach that mixed ecological and socio-economic criteria: it selected five river basins with a focus on farmland along rivers; it targeted administrative areas with high levels of poverty; it established a specific target area that worked with communities within a 10 km buffer zone of the Mount Kenya Forest Reserve and National Park (about 800,000 people); it used an ecosystem approach to identify Focal Development Areas (FDAs) within a 5 kilometer buffer zone of critical rivers (and other environmental hot spots) to initiate community-driven, bottom-up interventions. • The program implementers sought to boost the livelihoods of thousands of poor families or families at risk of being poor and to improve the overall conditions of agricultural production, water, and natural resources and established specific short-term objectives for all three project components that were achievable.
Multiple scales	• The program aimed at achieving outcomes at household level, community level, and landscape level and took into account the scale issues related to ecological and hydro-geological processes (for example, biodiversity habitat, wildlife corridors, wildlife conflicts, upstream-downstream linkages). Implementation of the program, including institutional strengthening, operated at multiple scales, either overlapping or nested within each other such as: very small Focal Development Areas to form Community Interest Groups, watersheds to organize water users, sub-basins to manage rivers and organize multiple Watershed Resource Users' Associations (WRUAs), forest zones or hilltop areas to restore degraded lands, and administrative areas to build and expand water and road infrastructure.
Multi-functionality	• The program was multi-functional and encompassed multiple land uses including food, cash crop, and timber production, areas important for wildlife, biodiversity, and hydrological services. Landscape interventions covered various livelihood systems and agricultural production systems (for example, based on mix cropping systems, livestock, or other natural resources). The program partners worked across government sectors including agriculture, water, and wildlife and included government institutions at national and at local levels such as the Water Resource Management Authority (WRMA) and WRUAs. Specific program interventions had to take into account and manage trade-offs between upstream and downstream stakeholders, between the need for poverty reduction (income generation) and safeguarding the environment, and between farm-level and off-farm-level impacts.

table continues next page

Table 5.8 How Were Principles of Good Practice for Integrated Landscape Management Addressed in the Kenya Case Study? *(continued)*

Principle of good practice	Explanation and examples

	• Landscape actions covered all major components of the programs and included:
	  – Support and build local institutions for sustainable management of natural resources such as WRUAs and Community Forest Associations (CFAs), for example by strengthening participatory planning procedures to develop and implement plans for sustainable water and natural resources management.
	  – Establish adaptive research and demonstrations to improve soil fertility and prevent erosion (for example, with the help of on-farm trials and demonstrations, soil fertility enhancement, seed multiplication and distribution for improved crop varieties).
	
	  – Adopt income-generating activities and create small Common Interest Groups (CIGs) with about 20–30 members. The income-generating activities were designed to be beneficial to the household and the wider community, especially downstream water users.
	  – Establish soil and water conservation and agroforestry practices beneficial to both farmers and downstream stakeholders.
	  – Strengthen watershed institutions (for example, WRUAs to prepare sub-catchment management plans), improve access to water resources for domestic and productive purposes, promote water-saving irrigation technologies and reduce point sources of water pollution.
	  – Strengthen community groups to improve their management of agricultural and forest lands (for example, rehabilitation of degraded forest reserves; efficient use of fuelwood; efforts to reduce human-wildlife conflicts with the help of fencing).
	  – Promote soil and water conservation on farmlands through on-farm demonstrations, field days, farmer-to-farmer extension, and the provision of matching grants.

Adaptive planning and management	
Adaptive planning and management and continual learning	• This was only achieved within small-geographic intervention areas and at more local scales.
Participatory and user-friendly monitoring	• Monitoring was set up to implement program inputs and outputs and monitor change at household and landscape levels for selected indicators. The program made investments to monitor water quantity (for example, setting up river gauges) and quality (short-term tracking during program implementation), which are expected to be fully integrated into monitoring efforts of the water sector over the long term.
Resilience	• Household resilience increased through new income-generating activities and higher incomes. Resilience of ecosystems supplying ecosystem services or a less attenuated delivery of water over the dry and wet seasons was probably achieved but not specifically tracked (one reason being that only core areas within five sub-catchment participated). Downstream users had in theory higher water supply during the dry season and less sediment positively affecting hydropower generation and drinking water supplies.
Collaborative action and comprehensive stakeholder engagement	
Multiple stakeholders	• The program engaged international, national, and local institutions such as the Global Environment Facility (GEF), IFAD, and Government of Kenya institutions at national and local scales. It involved beneficiaries of the program and their participation with various institutions such as: WRUAs; small Common Interest Groups (CIGs) with 20–30 members to develop local natural resource management plans and launch income-generating activities; and Community Forest Associations (CFAs). The program involved both public and private actors (for example, use of private sector service providers with a comparative advantage). Within the very local geographic intervention areas, the program sought to achieve multiple benefits at farm and landscape scales.

table continues next page

Table 5.8 How Were Principles of Good Practice for Integrated Landscape Management Addressed in the Kenya Case Study? *(continued)*

Principle of good practice	Explanation and examples
Negotiated and transparent change logic	• The stakeholders agreed on a process for actions to achieve goals and develop community projects. Project plans were developed in a participatory manner, the landscape-level implications of certain land and water management practices were discussed, and changes were made. Project actions focused on improving efficiency of current water, land, soil, and energy use and select interventions that avoided (or reduced) negative off-farm externalities. There was a negotiated and transparent change logic associated with the activities only within each geographic focal area. There was none at the watershed scale, since the program had not established an explicit process between upstream and downstream users (for example, upstream farmers in high-potential areas versus downstream irrigators, pastoralists, hydropower generators).
Clarification of rights and responsibilities	• The program used various targeting approaches (for example, specific targeting of disadvantaged groups, self-targeting) and mostly relied on an incentive system and demonstration efforts to motivate community members. Roles of contributions and the expected benefits were clear for participants. The program assessment reports provide no information on how conflicts were resolved.
Strengthened stakeholder capacity	• The community driven approach provided a platform for dialogue, sharing experiences, and learning. Capacity building and strengthening social, financial, and other capabilities were important MKEPP objectives.

to show measurable impacts at landscape scale. However, the following achievements were reported in the evaluation reports:

• *Water pollution reduced and riparian areas improved.* About 265 kilometers of riparian areas were stabilized by tree planting, mostly on farmland. The program reported a reduction in water pollution and silt levels for monitored sites.
• *Tree cover increased, carbon sequestered, and degraded hilltop areas rehabilitated.* About 5.25 million planted trees survived. Tree cover on agricultural land of participating farmers increased and selected hilltops were rehabilitated in the pilot area.
• *Improved hydrological monitoring system.* The program established 24 new river gauging stations and rehabilitated 54 stations. This investment in data collection infrastructure should improve analysis and associated information products (for example, to determine water allocation and provide early warning for drought and flood conditions).

Data on the basin-wide water quantity and quality effects are not readily available from MKEPP project assessment reports and not all landscape-level effects were monitored. Some farm-level soil and water management interventions require multiple years of operation to detect basin-level changes. It is difficult to attribute changes in water flows or water quality to specific interventions in selected Focal Development Areas (5-kilometer buffer zones along critical river sections)—too much of the "watershed was missing," and pilot program activities needed to be scaled to a larger proportion of the upper Tana River basin.

However, IFAD in collaboration with the World Data Center for Soils (ISRIC), the Kenya Agricultural Research Institute, and other research institutes carried out very detailed technical and economic studies and comprehensive stakeholder engagement to explore the feasibility of Green Water Credits (see box 5.1) for the whole upper Tana River basin (see technical documents and presentation at the Green Water Credits site). These studies provide an overview of the economic magnitude of the landscape-level benefits, although the presented economic calculations were not explicitly made for the changes in soil and water management practices that occurred for the farmers involved in MKEPP. For the upper Tana, potential watershed benefits are large in size and benefit-cost ratios are very favorable, once the supplier of

---

Box 5.1   Green Water Credit Scheme

A Green Water Credit is a financial mechanism (Payment for Ecosystem Services scheme) that links large downstream water users to the soil and water conservation efforts of thousands of upstream farmers. The idea is to reduce erosive farming practices upstream on smallholder farms, as was successfully piloted in MKEPP. Conservation measures such as mulching, bench terraces and contour bunds reduce erosion, reduce runoff and recharge ground water (increase of water in soils = green water). Farmers applying these techniques in their field benefit directly from better soil moisture management which is reflected in increased yields and associated income gains (usually within a few cropping seasons). Downstream users are benefiting from more reliable base flows in rivers during the dry season (because of upstream ground water storage), which is beneficial for dryland irrigators, hydropower operators, pastoralists, and downstream ecosystems depending on a minimum environmental water flow. Reduced sediment downstream also reduces the operating costs to maintain hydropower reservoirs and run turbines.

A credit arrangement using public and commercial financing and payments from large downstream water users has been proposed to launch the Green Water Credit scheme. Sufficient and long-term financing is necessary to recoup the investment costs for constructing the soil and water conservation measures and provide funds for their maintenance. van Steenbergen, Tuinhof, and Knoop (2011) highlighted other sources of funding for successful soil and water conservation efforts:

- *Public funds and subsidies.* Safety net programs, such as discussed in the Tigray case study (MERET) could be such a source.
- *Spare labor and drawing dividends from existing natural capital.* The soil and water conservation efforts supporting regreening in Niger were achieved with off-season labor that had very low opportunity costs. For the farmer-managed natural regeneration, there was no need to buy inputs since the trees regenerated naturally from their already established roots.
- *Off-farm income.* In the 1980–90s, farmers in Machakos County, Kenya funded their *fanya juu* terraces with off-farm income. Remittances of migrant workers were the source of funding for construction of earthen storage dams in the Republic of Yemen.

Sources: van Steenbergen, Tuinhof, and Knoop 2011; Tiffen, Mortimore, and Gichuki 1994; Mortimore, Tiffen, and Gichuki 1993; Pye-Smith 2013.

---

the ecosystem services can be linked to their downstream users (using a landscape-encompassing mechanism).

Figure 5.2 summarizes the major costs and benefits of improved soil and water management in the upper Tana River basin. The following assumptions were made for the economic valuation:

- *Upstream farmers (water suppliers)*
  - ○ 40,000 square kilometers intervention area including 150,000 smallholders with improved soil and water management practices.
  - ○ This represents about 20 percent of the smallholder farmers, the minimum investment required to have major impact on sedimentation. If 50 percent of the smallholder farmers adopt improved soil and water conservation up-stream, there would be a considerable impact on water quantity downstream.

- *Downstream water users*
  - ○ Kenya Electricity Generating Company (KenGen)—5 dams provide about 80 percent of Kenya's electricity.
  - ○ Irrigators (for example, Kakuzi, Del Monte, and Yatta canal dryland farm-ers): currently 7,000 square kilometers planned to increase to 20,000 square kilometers.
  - ○ Nairobi City Water and Sewerage Company.
  - ○ These major users determined the value of reduced sediment based on their operating and other costs.

The downstream benefits of better soil and water management in the uplands are far greater than the costs: annual benefits range between US$12 and US$95 million; annual costs range between US$2 and US$20 million depending on which soil and water measures are implemented and the level of adoption among farmers.

Assuming a 20 percent adoption scenario, the annual water benefits would be US$6–48 million compared with costs of US$0.5–4.3 million (Green Water Credit website; van Steenbergen, Tuinhof, and Knoop 2011). Half of these ben-efits would come from the decreased silting up of hydropower reservoir. Benefits such as higher crop yields, flood mitigation, carbon sequestered in soils, or dam-ages avoided in coastal ecosystems are not included in this economic estimate. The Green Water Credit scheme will require start up financing of US$50 million for the first five years to establish soil and water conservation measures, with no downstream benefits (initial investment benefits will increase from Year 4).

Insights and Lessons Learned

The two IFAD programs and the proposed Green Water Credit scheme for the upper Tana River provide insights on the applicability of the principles of good practice for improved landscape management. As in the previous two case stud-ies for Ethiopia and Niger, the Kenya example focused on enhancing farmers' livelihoods, promoting community engagement, and implementing activities within multiple sectors and at multiple scales.

Figure 5.2 Upper Tana River Green Water Credits: Costs and Benefits

Upstream

Benefits
Extra water
Less soil erosion
Improved food security
Higher yields/ha
Higher net worth of land

Costs
Education/Training
Equipment
Community building
Legal

Downstream

Benefits
Extra water
Improved public water supply levels
Improved operational liability
Better public services
Lower maintenance costs
Lower insurance rates
Less flooding and damages

Costs and benefits at basin level

Upstream on-site benefits	without (million m³)	with (million m³)
Groundwater recharge	5,960	13,860
Soil loss	25,600	13,150
Surface runoff	4,460	1,860
Crop transpiration	4,450	9,850
Soil evaporation	4,050	160

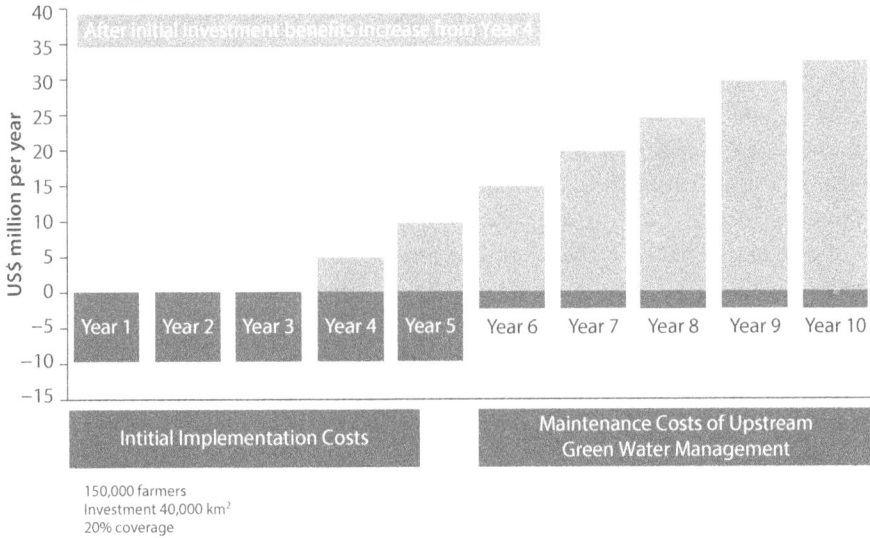

Downstream on-site benefits	without (million m³)	with (million m³)
Reservoir inflow	5,194	10,372
Sediment inflow	3,117	1,241

After initial investment benefits increase from Year 4

US$ million per year

Year 1 Year 2 Year 3 Year 4 Year 5 Year 6 Year 7 Year 8 Year 9 Year 10

Intitial Implementation Costs

Maintenance Costs of Upstream Green Water Management

150,000 farmers
Investment 40,000 km²
20% coverage

Source: Onduru and Muchena 2011; Muchena et al. 2011; van Steenbergen et al. 2011; Porras, Grieg-Gran, and Meijerink 2007; and Green Water Credits Kenya site.

It turned out to be difficult to implement an ecosystem-based approach (using Focal Development Areas as an entry point to organize activities) and work across government sectors.

• Government staff from different ministries had trouble working cross-sectorally without a specific mandate or authority to collaborate.

- Rivers are often the boundary between administrative areas—a 5-kilometer buffer zone on each side of the river for the selected Focal Development Area sometimes required coordination across multiple administrative units.
- In MKEPP, some program activities were too focused within the protected area, leaving out the farmland area, and some environmental conservation efforts were at the expense of rural livelihoods (a result of blending two different funding sources—GEF and IFAD—with different mandates focusing on biodiversity conservation and rural livelihood improvements, respectively).

The pilot program selected project activities with a focus on efficiency of water, land, soil, and energy use (which generally have positive externalities at landscape scale) and avoided activities with negative off-farm externalities. No specific institution was created to facilitate landscape management for the whole upper Tana ecosystem. Instead, the pilot program enhanced the capacity of existing water sector institutions at various scales.

There was not an explicit process between upstream and downstream users within the pilot program (for example, upstream farmers in high-potential areas versus downstream irrigators, pastoralists, hydropower generators). The pilot program supported local and regional WRUAs and collaboration with water institutions at watershed and basin-level scales to establish the general conditions for better watershed and basin-scale water management. A comprehensive water sector and policy reform had been completed in Kenya. These institutions have the mandate to assess upstream and downstream linkages and negotiate synergies and trade-offs associated with certain upstream farming practices or landscape restoration efforts.

Drylands are linked to high-potential areas and this can be used to increase resilience.

Changes in land and water management in high-potential areas of Mount Kenya can improve the situation for dryland irrigators.

Analyzing and designing a program at landscape scale can create new opportunities.

A landscape perspective and executing a Green Water Credit effort at landscape scale can save costs for hydropower companies and municipal water supply companies, create revenues for upland farmers (and improve their yields and resilience), and improve water supply and resilience for downstream irrigators in the drylands. In the case of the upper Tana, every dollar invested by an upland farmer in specific soil and water management practices provides about 10 dollars of benefits to downstream water users (in addition to the on-farm production benefits).

A bottom-up approach using community-level institutions with an ecosystem perspective can become the foundation for more comprehensive landscape-wide interventions.

MKEPP demonstrated the usefulness of combining community-driven approaches with watershed and ecosystem targeting, but was too small and fragmented in geographic scope to achieve large measurable landscape-level impacts (changing hydrological flows). It created, however, greater community capacity and short-term livelihood gains for farmers, which are a strong foundation for comprehensive scaling up to the whole upper Tana basin (UTaNRMP) and for developing a more ambitious integrated landscape-level intervention (such as a Green Water Credit scheme).

Insights from the Case Studies That Are Relevant to Implement Integrated Landscape Management

How Were Principles of Good Practice Applied across All Three Case Studies?

The three case studies demonstrate that integrated landscape management incorporates and builds on past approaches (for example, integrated watershed management) and seeks to scale more sustainable management practices at farm level (for example, water harvesting, FMNR of trees). The 10 principles of good practice represent a flexible framework for scaling investment at landscape scale (see side-by-side comparison of three case studies in table 5.9). Not all principles need to be applied in a comprehensive manner at once. The principles can be sequenced and adapted to local conditions.

Both the Ethiopia and Kenya cases started out as development cooperation programs focused on food security, livelihood development, and more sustainable land, water, or natural resources management practices. The respective programs incorporated a spatial approach to plan, prioritize, and execute project interventions. Initial programs evolved into new programs over time from pilot efforts or a focus on a single sector such as water. Not all integrated landscape initiatives, however, need to be implemented following such a program planning approach. The Niger case suggests that a useful entry point for integrated landscape management can be a general concern about soil fertility and low yields, followed by a systematic search for farm innovators that have restored farm productivity or degraded lands, which is then scaled with the help of sustained support for short-term training and peer-to-peer learning in combination with targeted interventions aimed at improving the policy and regulatory framework and other enabling conditions for widespread adoption. Once a sufficient number of farmers have adopted new management practices such as FMNR, landscape-level goals and additional interventions to safeguard these investments in trees can be introduced and help to expand and accelerate scaling of these bottom-up efforts.

Summing Up: Key Points for Consideration in Developing Integrated Landscape Management Interventions

The following key points and insights gained from the case studies are relevant and important to keep in mind when developing new integrated landscape management interventions.

Table 5.9 How Were Principles of Good Practice for Integrated Landscape Management Applied in the Three Case Studies?

	Ethiopia	Niger	Kenya
Landscape goal(s) and multiple objectives			
Common concern entry point	• Food security and safety net • Livelihood development and rural infrastructure building • Watershed restoration (landscape goal)	• General concern about soil fertility and yields • No specific landscape goal and objective for FMNR • Village/landscape goals established later to safeguard trees and other farmer investments	• Improve rural livelihoods and sustainable management of water, agricultural lands, and ecosystem services • Geographic targeting within the upper Tana River
Multiple scales	• Implemented at farm level, planned at watershed • Improve farm productivity and income diversification • Watershed benefits (exclosures)	• Benefits at farm-level and within farming system • Collaboration at village and inter-village scales • Support organizations worked through networks	• Outcomes at household, community, and landscape levels with implementation within selected riparian areas, watersheds, sub-basins and forest zones
Multi-functionality	• Crop producers and pastoralists • Multiple types of land use in watershed	• Higher crop yields, increased fodder and firewood • Decentralized resource management across the landscape • Forest service used "hands-off" approach related to trees on farms (and did not enforce restrictive rules)	• Multiple land uses, various livelihood systems and agricultural production systems, and biodiversity • Decide on trade-offs: farm versus off-farm/down-stream
Adaptive planning and management			
Adaptive planning and manage-ment and con-tinual learning	• Within watersheds	• Farmers and rural communi-ties proceeded in step-wise process: started with chang-es in farming practices and then built local institutions to manage and safeguard their farm-level investment in trees	• Within small geographic areas (not within watershed)
Participatory and user-friendly monitoring	• MERET trained communi-ties in measuring results • No collection of baseline data for MERET • Overall, programs in Tigray could be much stronger	• Apart from farmers monitor-ing within their fields, only occasional studies	• Monitoring of program inputs and outputs • Monitoring of changes at household level and at landscape level (water quality)
Resilience	• WFP included disaster risk reduction measures • MERET includes liveli-hood diversification components • Improved food security	• Growing evidence that farmer-managed natural regeneration practices are contributing to resilience	• Increased household resil-ience through new income-generating activities and higher incomes • Resilience of ecosystems supplying ecosystem services

table continues next page

Table 5.9 How Were Principles of Good Practice for Integrated Landscape Management Applied in the Three Case Studies? *(continued)*

	Ethiopia	Niger	Kenya
Collaborative action and comprehensive stakeholder engagement			
Multiple stakeholders	• Community resource users groups • Administered by MoA, implemented bottom-up with *Woreda*, *Kebele*, and watershed development committees	• Smallholder farmers are key stakeholders • Village-based organizations play key role in peer-to-peer learning and adopting and enforcing local rules	• Government institutions at national and local scales • Water resource users' associations, common interest groups and private sector service providers
Negotiated and transparent change logic	• Agree to restore productivity of degraded land with specific actions on 'private' and 'public' land • Exclosures for steep hills	• A majority of farmers agreed to change resource management practices to restore productivity of degraded land • Once trees regenerated, community established rules	• Community projects improved efficiency of water, land, soil, and energy use and select interventions that avoided (or reduced) negative externalities
Clarification of rights and responsibilities	• Responsibilities for various restoration efforts • Registration (land certificates) within project site (SLMP) • Experiment for restored community land delivered to landless youth	• Rights on private farmland and rules within community • Government policies were supportive of the ownership of trees on farms and giving management rights to farmers	• Self-targeting and targeting of disadvantaged groups • Incentives and demonstration efforts to participate • Contributions and benefits for community projects were clear (no specific efforts on land/water rights in general)
Strengthened stakeholder capacity	• Capacity building as part of MERET • Creation of new institutions (watershed committees)	• Assistance with farmer-to-farmer visits, short-term training, and peer-to-peer learning	• Community driven approach provided platform for dialogue, sharing experiences and learning • Capacity building was a MKEPP program objective
	PRINCIPLE COMPREHENSIVELY APPLIED	PRINCIPLE PARTIALLY APPLIED	PRINCIPLE MINIMALLY APPLIED

Note: FMNR = farmer-managed natural regeneration; MERET = Managing Environmental Resources to Enable Transitions Program; MoA = Ministry of Agriculture; SLMP = Sustainable Land Management Program; MKEPP = Mount Kenya East Pilot Program.

Ethiopia

Early efforts to restore drylands and address rural poverty and food insecurity were top-down, focused narrowly on technical aspects of hydrological improvements, and selected watersheds that were too large in size. As a result, many of these early projects did not achieve the expected results and failed to encourage long-term participation and maintenance of restoration interventions. Poorly designed technical watershed restoration efforts that did not understand the hydrology of an area, disregarded the rights and ownership of farmers and communities, and lacked a focus on long-term maintenance and training of land users were eventually seen as failures in meeting long-term national and regional goals to restore drylands.

- Over time, more recent programs learned from shortcomings of previous ones. An increasing number of integrated landscape management principles of good practice were incorporated, resulting in greater adoption rates, benefits to farmers, benefits at watershed and landscape scale, and increased likelihood of long-term sustainability. These experiences validate many of the outlined principles of good practice for integrated landscape management. Ethiopia's local-level participatory planning approach guidelines and guidelines for Community-Based Participatory Watershed Development overlap considerably with these principles and could be instructive for other dryland areas that are mountainous and include all three major aridity zones, such as highlands in eastern and southern Africa.
- Over the past four decades, at least three major programs have focused on addressing food security and drylands restoration in Tigray: Project Ethiopia 2488, which eventually evolved into the Managing Environmental Resources to Enable Transition (MERET); the Productive Safety Net Programme (PSNP); and the Sustainable Land Management Program (SLMP).
- The entry points for these programs were disaster relief, social safety nets, and drylands restoration, which provided the initial financing plus labor contributions from participating communities.
- These programs focused on building on- and off-farm assets and used a spatial approach, working at the watershed level but promoting farm- and household-level participation to implement interventions. Farmers invested in farm-level efforts to decrease runoff and capture more of the rainfall. These efforts were coordinated across watersheds and communities within these watersheds. Communities also worked to restore marginal lands including steep slopes and degraded hilltops with the help of exclosures and assisted natural regeneration of shrubs and trees. The programs also focused on social and economic interventions including investments in roads, irrigation systems, schools, and clinics and provision of agricultural inputs (for example, fertilizers) to improve market access and improve human health and education.
- Cash- or food-for-work interventions were helpful for creating social safety nets that allowed expansion of sustainable land management interventions and increased farm-level awareness of ecosystem restoration and the importance of interventions such as exclosures.
- Reviews of sustainable land management efforts and integrated landscape management programs in Ethiopia and beyond are showing that a longer time horizon is required to get positive returns from investments in sustainable land management practices at landscape scale, as ecosystem health must first be restored to enable improved land production.
- The transition in Ethiopia from centralized efforts to address land degradation (under the Derg regime) towards a decentralized approach has helped to

improve ownership over land by rural users and improved maintenance of sustainable land management interventions. The focus of the Sustainable Land Management Program (SLMP) on supporting land tenure reforms provides an additional incentive for farmers to participate and maintain sustainable land management investments.

- The programs now promote interaction of a variety of government actors at different scales. For example, SLMP and MERET promote community implementation of interventions, but communities also interact with local, regional, and national government actors.

- Programs in Ethiopia (namely MERET) are beginning to transition from targeting the poorest households to systematically targeting communities and watersheds at scales that promote ecosystem and community connectivity. Moreover, MERET is beginning to move past social safety nets towards providing more technical assistance for livelihoods diversification and building savings and credit. The past four decades have helped to establish a new foundation for planning by creating more informed and empowered communities to participate in future land-use planning and decision making. However, there are still gaps in dialogue across multiple watersheds and sub-basins that should be addressed.

- Exclosures have a significant potential for scaling up in the densely populated highlands of Ethiopia. They produce multiple ecosystem benefits (for instance, reduction of erosion and runoff, groundwater recharge).

- Monitoring and evaluation systems for programs continue to be weak, leading to a lack of quantitative evidence on how human welfare is improving and how well sustainable land management interventions are working at different scales.

The investments in land and water management in Tigray have led to an increase in the number of smallholder farmers able to cope with external shocks like drought. During the last two decades the physical capital has increased significantly as an estimated 960,000 hectares or almost 25 percent of the total region have been treated with one or more conservation techniques. The natural capital has been expanded as exclosures have increased vegetation cover and biomass. The human capital has grown because interventions have built the skills of smallholders. Even though many smallholders are still poor and vulnerable, many perceive a reduction in poverty (Sutter et al. 2012). Social capital building is reflected by stronger and empowered village-level organizations, which have improved technical and organizational skills.

Niger

- Local investments in the improved management of land and water resources have proved to be an effective entry point for achieving landscape-level benefits for rural communities. The widespread adoption of FMNR and water harvesting practices has restored soil organic matter, reduced rainfall runoff, and increased soil moisture available for crops, as well as enabled the more efficient use of fertilizers and the diversification of crops produced in both the rainy and dry seasons. Trees in agroforestry parklands reduce local temperatures, and people, crops and livestock are all better protected from the effects of wind erosion and sand-storms. Farmers can rely on a broader range of goods and services produced by more complex farming systems and because of this, rural communities are better buffered from the impacts of erratic rainfall.

- In the Kantche Department of Niger near Zinder, where there is now a higher density of on-farm trees following the widespread adoption of FMNR, there have been grain surpluses in every year since 2007, ranging from 13,818 tons in 2011 to 64,208 tons in 2010, despite population increases and variable rainfall.[5] In addition to benefiting from higher crop yields and grain surpluses linked to the increased density of trees on farms, rural communities in this region are able to harvest and sell poles, fodder, firewood, edible leaves, and other products that provide additional or alternative sources of income even when the production of rain-fed crops is reduced.

- Enabling farmers to address the root causes of land degradation, such as declining soil fertility and loss of resilience of rural production systems has been critically important. Widespread adoption of FMNR and other improved management practices has been driven mainly by the short-term benefits that accrue to farmers who invest in the improved management of trees on farms and in managing soil fertility and rainfall. Investments by NGOs and others in identifying successful practices, and in supporting farmer-to-farmer visits, training, outreach, and communication have facilitated the emergence of farmer innovators and local champions for FMNR and other improved practices. And while NGOs played a catalytic role in triggering the landscape-level transformations, support by the national government and others for needed policy and institutional reforms that addressed the barriers to widespread adoption of these improved practices was also very important. Farmers were empowered by these reforms, which clarified and secured their rights to manage trees on farms and enabled community-based organizations to adopt and enforce rules aimed at protecting the investments of farmers in restoring agroforestry parklands.

Kenya

- Blending of bottom-up and top-down approaches can achieve results at household and landscape scales. The case study demonstrates that it is possible to plan a program at watershed or ecosystem scale to identify focal areas. There, bottom-up, community-driven activities can be selected and implemented that deliver livelihood gains and other benefits at household and community levels and also provide positive externalities for downstream water users or other ecosystems or sectors (for example, forests, wildlife, biodiversity).

- Working across sectors (for example, water, agriculture, wildlife, forests) and across administrative areas is possible, but not easy. Funding modalities and institutional mandates can discourage collaboration. Some geographic areas may be optimal in delivering watershed or ecosystem benefits, but are difficult to coordinate administratively, requiring a local compromise to make implementation feasible.

- A dryland area may depend on ecosystem services from other areas and may provide ecosystem services for other areas. A systematic assessment of ecosystem service provision and dependence in drylands (for example, open space for wildlife and livestock, water-related services, carbon storage) can be a useful exercise and could create new funding streams or other rewards (for example, more secure use rights) for households or communities.

- Increasing the geographic scope of a program can achieve landscape-level scale effects, engage a new group of stakeholders, and create new opportunities for financing or solving other development challenges. The program (MKEPP) moved beyond a household-level perspective (for example, improved land and natural resource management practices) and sought additional outcomes at community and watershed level. MKEPP worked with and strengthened existing institutions at different scales and within different sectors, which was a practical approach and feasible within the given budget and time horizon of the program. It established the foundation for future scaling and more comprehensive landscape-wide interventions (for example, Green Water Credit schemes).

- A landscape approach could align the interests of less powerful water users with major water users and result in a more reliable and fair water distribution. Within the upper Tana River, the community-based smallholder schemes along the Yatta canal were at the tail-end of water supplies and among the less influential water users. A Green Water Credit scheme (or other arrangement) involving all major water users could establish a long-term mechanism to improve supplies and ensure a fair distribution.

- An integrated landscape management initiative could explore how investment in improved soil and water management could be aligned with support that improves farm outputs, farm returns, and market access. Support for improved crop and livestock varieties and market access could be aligned with improved soil and water conservation investments and shift the price ratio of farm outputs and inputs, making soil and water conservation practices more profitable and boost farm (private) returns and public good provision.

- A sequential step-by-step approach may be a sensible way to advance integrated landscape management. Start out with activities that provide relatively quick benefits for farmers and land users, create positive externalities for other sectors, ecosystems or downstream beneficiaries, and do not create negative externalities (as was done in the pilot program); strengthen the capacity of existing institutions that have a mandate to plan, facilitate, or coordinate land-scape-level interventions, assuming that these institutions are effective (for example, water sector institutions in Kenya); only introduce more complex landscape-level interventions (for example, Green Water Credit schemes) once initial activities have been scaled, stronger community capacity and short-term livelihood gains for farmers and other households have been achieved, and a clear demand for complex landscape-level intervention has been expressed.

Lessons Learned for Implementing Integrated Landscape Management

The ecological and economic evidence from these case studies shows that integrated landscape management can enhance efforts to invest in tree-based systems and improved livestock management and support productivity increases for rain-fed cropping. Integrated landscape management efforts have helped to coordinate the actions of multiple land users and other stakeholders, reduced conflicts, and improved overall governance of water, land, and other resources. The following observations from the case studies are relevant when new integrated landscape management initiatives are being developed:

- There is no blueprint on how to implement integrated landscape management.
- The case studies confirm the relevance of the 10 principles of good practice, but more experience is required to identify the best sequencing and phasing of steps for effective implementation of integrated landscape management in specific locations and farming systems. Such information is currently being collected by the Landscape for People, Food and Nature initiative.
- Integrated landscape management requires both a long-term perspective (greater than 10 years) for many landscape-level effects to take place and an emphasis on promoting practices that provide quick short-term (one to three years) benefits at household level to motivate participation and support by farmers, herders, and communities.
- Case studies illustrate the need for long-term monitoring and evaluation aspects to be included in programs to understand long-term resilience to climate change, and other stressors and landscape-level benefits. While the three case studies were partially chosen because they had better data and were better studied than other case studies, they still lack comprehensive long-term monitoring and evaluation components.
- Integrated landscape management initiatives can be launched with the help of short-term programs (for example, five to seven years) as long as they are designed with a landscape-level vision in mind and as long as they learn systematically from successes and shortcoming of past programs.
- A mix of participatory bottom-up approaches with a landscape vision can provide a useful framework to launch an integrated landscape management initiative. This can be further strengthened by systematic analyses of interlinkages of

ecosystems, production systems, and the needs of different land users at landscape level (for example, mobility for pastoral livestock keepers, corridors to encourage wildlife migration, protection of riparian corridors to safeguard water quality).

• For areas with severe degradation, it might be necessary to invest first in sustainable land and water management practices covering social, technical, and environmental aspects to regenerate ecosystem health (for example, livestock exclosures, grazing bans, rainwater harvesting infrastructure, education and skills development for primary stakeholders, social safety net programs to provide food and cash for villager implementation of interventions). Such programs should take advantage of low-cost options (such as self-scaling farmer-led restoration efforts) and be tailored to the local context which is based not only on who is involved, but also the farming system classification.

• When launching new integrated landscape management programs, sufficient resources must be available to support adaptive planning and management at landscape scale. Monitoring and evaluation systems need to be designed to provide quantitative evidence on how project interventions are achieving results at various scales (for example, household, community, landscape).

Integrated landscape management should not be seen as a "one-size-fits-all" solution but rather a flexible framework for scaling investments at a landscape level to maximize ecological, economic, and social synergies and minimize negative trade-offs.

Notes

1. An exclosure is a demarcated area established to restrict grazing and wood harvesting and allow for regeneration of vegetation (Meire et al. 2012).
2. http://hdr.undp.org/sites/default/files/hdr2013_en_summary.pdf.
3. See http://www.wri.org/publication/improving-land-and-water-management.
4. See http://www.esteri.it/MAE/doc/6_40_175_f.pdf and http://www.fao.org/docrep/x3989e/x3989e05.htm.
5. Source: Data compiled by Yamba and Sambo 2012, from the National Committee for the Prevention and Management of Food Crises and FEWS, and cited by C. Reij 2013 in Climate-Smart Agriculture, Food Security and Water In Africa's Drylands: Lessons from Experience (presentation at WRI).

CHAPTER 6

Recommended Policies and Other Interventions to Advance Integrated Landscape Management and Enhance Resilience in Drylands

This book provides theoretical and case example evidence of the multiple economic, ecological, and social benefits possible through integrated landscape management. The previous chapters document a variety of integrated landscape management initiatives and some key principles recommended to make it work. In some regions, development and restoration policies have evolved to adopt more of these principles. The Ethiopia case study, for example, highlights how transitioning from a sectoral, top-down, and centralized approach to a multi-sectoral, participatory, and decentralized approach helped to enhance and scale up sustainable land and water management and generate both on-site farm-level and off-site landscape-level benefits. In other regions, new programs are emerging, bringing innovative concepts and new actions to the table while promoting integrated landscape management principles.

The review of ecological and economic evidence and detailed case examples show that integrated landscape management can enhance efforts to invest in tree-based systems and improved livestock management, and support productivity increases for rain-fed cropping. Integrated landscape management efforts have helped to coordinate the actions of multiple land users and other stakeholders, reduced conflicts, and improved overall governance of water, land, and other resources. Integrated landscape management is thus a useful approach to enhance the intensification of drylands cropping systems and will, in many locations (but not always), result in multiple wins including the following: improved farm productivity, water benefits at farm and landscape level, carbon sequestration, biodiversity and other ecosystem services benefits, and higher climate resilience.

Based on evidence gathered in this book this chapter discusses policy recommendations and related interventions that can be used to trigger and accelerate the scaling up of these benefits through integrated landscape management across Sub-Saharan African drylands to restore and increase household and ecological resilience. Policies are needed to develop the framework conditions necessary to both initiate new programs and modify and scale up existing restoration and resilience efforts in Sub-Saharan African drylands. The recommendations cover a variety of policy instruments (for example, technical assistance, knowledge-sharing, economic incentives, regulatory instruments) to address the following important challenges for advancing integrated landscape management to enhance resilience in Africa's drylands, including:

- Lack of knowledge and awareness about integrated landscape management within national and local governments, private sector, and civil society actors.
- Institutional barriers that impede addressing complexities at the landscape level.
- Poor availability of and access to location-specific data about land, water, and natural resource use and limited local planning capacity to optimize land use within landscapes.
- Difficulty in ensuring management of trade-offs and provision of adequate incentives for needed behavioral changes and sustainability.
- Fragmented financing for drylands restoration and integrated landscape management.

To address these important challenges and advance integrated landscape management, recommendations cover six broad areas:

1. Clarify land rights and responsibilities
2. Encourage multi-stakeholder involvement and collective action
3. Overcome institutional barriers to integrated landscape management
4. Create conditions for adaptive planning and management
5. Create mechanisms and supporting policies for sustainable and long-term financing of integrated landscape management
6. Invest in a solid evidence base and knowledge-sharing platforms for integrated landscape management

It should be noted that these recommendations are not intended as a solution set, but rather as policy options that can be explored and applied to the local context. The final examples show how these general policy recommendations can then be further tailored depending on agro-ecological zone.

1. *Clarify land rights and responsibilities*. Primary stakeholders involved in resto-ration efforts need the security and certainty to invest physically and finan-cially in improved land and water management practices and supply public goods to achieve landscape-level benefits. This requires clarifying land rights and responsibilities through land tenure reforms and reforming land-use

codes to give communities access to resources and a sense of ownership over the future of that land. Policy options and actions to support this include:

- Decentralize policies for natural resources management and provide more authority to community resource organizations to empower them to make decisions.
- Reform land-use planning and land tenure policies. For example, in Niger, a reform of the Forest Code, land tenure reforms, and popularization of the Rural Code led to an increased sense of ownership of farmers over trees on their property and increased regeneration of trees, which led to improved soil productivity and ultimately agricultural productivity.
- Support land titles with legal validity and transparency.
- Modernize legal systems if necessary to enable local and national governments to incorporate flexibility into land tenure systems that consider management of landscape-scale benefits (for example, use of common pool resources, resilience of pastoral production systems) and to enhance transparency within land-use policies.

2. *Encourage multi-stakeholder involvement and collective action.* Integrated landscape management affects stakeholders in different ways, and trade-offs need to be well managed by all concerned stakeholders. To encourage participation across sectors and stakeholders, it will be necessary to identify winners, compensate losers, and increase incentives to encourage adoption of integrated landscape management through innovative incentives and risk reduction measures (for more information see chapter 3). The planning and organization of interventions at the scale of a targeted landscape does not automatically lead to sustainability and needs to be carefully monitored. Multi-stakeholder involvement, prioritizing communities and primary stakeholders, can be supported by the following actions:

- Create or invest in incentive schemes to compensate losers and encourage their participation (for example, Payments for Ecosystem Services schemes).
- Create mechanisms or institutions to help with conflict resolution and manage trade-offs between groups representing different sectors, cultures, and interests.
- Create mandates and strengthen authority for government staff to work across ministries and government sectors.
- Prioritize full participation of primary stakeholders (for example, farmers, herders, producer groups) and marginalized stakeholders (for example, women, landless, youth) as a key entry point and foundation for stakeholder involvement in integrated landscape management design, implementation, and evaluation. This can be supported through decentralization policies and investments in community resource groups.

- Provide opportunities for private sector engagement as well as NGOs and other civil society organizations by including them in stakeholder discussions.
- Encourage policies that support framework conditions for collaborative action and comprehensive stakeholder engagement such as freedom of speech and assembly, and formation of and legal standing of producer associations and other common interest groups.

3. *Overcome institutional barriers to integrated landscape management.* A significant investment in institutional reforms and capacity building is needed, along with assessments of sector-specific mandates of different ministries to resolve the challenges of working across sectors. Reducing institutional barriers, redundancies, and conflicting messaging between government agencies to better target funding and communication and show government support for integrated landscape management are all needed, and can be supported with the following actions:

   - Conduct a thorough review, using impartial external stakeholders, of current restoration programs and policies: to identify barriers to scaling up restoration success stories and implementing integrated landscape management principles; to identify gaps in staffing and policies needed to promote integrated landscape management; and to better target funding for drylands restoration activities and programs.
   - Reform agricultural input subsidies to ensure a level playing field for both the provision of agricultural inputs and the improvement of agricultural management practices.
   - Reform land-use planning policies that stymie farmer and other land-user innovation related to land and water management practices.
   - Reform agriculture, forestry, and rural development policies that inhibit cross-sectoral collaboration among government agencies.
   - Advance national policies that promote resilient agricultural landscapes and encourage provision of public goods from agricultural landscapes.
   - Create a common set of guidelines for drylands restoration that incorporates principles of good practice for integrated landscape management with endorsement from relevant government agencies to show solidarity in promoting integrated landscape management. Adopt guidance into national climate change action plans (for example, India's National Watershed Development Guidelines).
   - Link government planning agencies and proponents of integrated landscape management to scientific and policy research on integrated landscape management (to reduce the science-to-policy gap) with a particular emphasis on raising awareness on the cost-effectiveness and co-benefits of integrated landscape management.

4. *Create conditions for adaptive planning and management.* For many dryland areas, local planners have very limited access to GIS (geographic information system) data of land cover, land use, water supply, natural resources use, and other information needed to identify the most promising areas for improved land and water management at a landscape scale and evaluate changes in human well-being resulting from integrated landscape management interventions. In addition, local planning capacity is generally low because of a persistent marginalization of drylands. Conditions for adaptive planning and management can be improved by the following:

- Provide guidance on how planning for integrated landscape management can be aligned with national, regional, sectoral, and local planning processes.
- Encourage spatial aspects of planning in local development planning and strengthen participatory land-use planning policies where they do not exist.
- Create incentives for more coordinated and systematic planning and linking government budget and planning (for example, link public budgets to land-use planning outcomes such as landscape-level provision of public goods).
- Invest in research opportunities and integration of long-term resilience monitoring and evaluation at the project scale to improve understanding of which interventions and combination of interventions are working and which are not.

5. *Create mechanisms and supporting policies for sustainable and long-term financing of integrated landscape management.* To help overcome initial upfront costs and support long-term maintenance of interventions and adaptive management, mechanisms and supporting policies for sustainable and long-term financing need to be created. Financing for drylands restoration or just sectoral investments are frequently spread across a multitude of government and other implementing agencies, resulting in programmatic and funding gaps and inefficiency. The following actions can provide a foundation for more sustainable and long-term financing:

- Support local investment in innovations to improve land and water management and enhance resilience.
- Create incentives and reduce perceived risks to encourage public and private investments through risk reduction guarantees and other risk reduction mechanisms.
- Identify opportunity costs (at household and landscape scales) associated with the adoption of sustainable land management and watershed protection activities as well as sustainable agricultural intensification practices.
- Assess the needs for initial government guarantees, subsidies, payments or other measures to overcome perceived risks and constraints, and to trigger and accelerate changes in behaviors and local investments in desired practices.

- Consider opportunities to align land tenure reforms and measures aimed at strengthening community land tenure and resource management rights with other mechanisms aimed at reducing risks and increasing incentives for targeted investments.
- Provide additional financing and innovative approaches to fill gaps and inefficiencies in financing of integrated landscape management programs.
- Work with Ministries of Finance, donors, or financiers to promote the consolidation of financing for mainstreaming improved land and water management. Reinforce this with increased financing for integrated landscape management in targeted dryland landscapes where rural households are particularly vulnerable and where there are significant opportunities for scaling agroforestry and other improved land and water management practices.
- Establish a land-use policy that provides fair compensation or new economic opportunities in case of land-use limitations for individuals, groups, or companies resulting from integrated landscape management.
- Create and reform policies to link the financing for integrated landscape management to the mobilization of needed investments in emergency response and safety net programs, food security and promotion of climate-smart agriculture, investments in climate change adaptation and mitigation, as well as poverty reduction.
- Support or create micro-finance schemes to promote small business development and market access infrastructure.
- Support or create revolving loans or maintenance funds to encourage long-term maintenance by primary stakeholders of sustainable land and water management interventions.

6. *Invest in creating a solid evidence base and knowledge-sharing platforms for integrated landscape management.* Detailed knowledge about integrated landscape management within national and local governments, private sector, and civil society actors could be greatly enhanced. It is therefore necessary to track what is working and what is not working, to reduce communication and dissemination costs for implementing agencies of restoration activities, to encourage innovation in improved land and water management, and to invest in creating a solid evidence base and knowledge-sharing platforms. The following actions can support this objective:

- Establish guidance and guidelines for monitoring and evaluation, recognizing the importance for long-term monitoring to account for adaptation to stressors (especially climate change). Monitoring and evaluation systems and guidelines should include principles of adaptive management.
- Invest in research institutes, academic extension programs, and NGOs to establish and track important restoration indicators, create knowledge-sharing platforms, and establish monitoring and evaluation systems that link to existing drought monitoring and early warning systems.

- Invest in peer-to-peer learning programs and increase recognition given to farmer and other land-user innovations.
- Invest in infrastructure needed to scale up integrated landscape management including communication support (for example, radio information, documentaries, agro-meteorology).
- Identify champions and leaders of integrated landscape management who can play a critical role in raising awareness and promoting this approach and represent different cultural and resource groups and sectors. These champions can be assisted by investing in opportunities to take the lead in documenting integrated landscape management successes.
- Invest in making location-specific data (for example, GIS maps, land and water use, natural resource use, agriculture and forestry statistics) more accessible to practitioners of integrated landscape management and monitoring and evaluation teams, to enhance the quality of baseline data, ensure tracking of drylands resilience indicators, and avoid maladaptation.
- Invest in developing training materials and source-books to provide information on the role and importance of adaptive management in integrated landscape management.

For areas with severe degradation, it might be necessary to invest first in sustainable land and water management practices covering social, technical, and environmental aspects to regenerate ecosystem health (for example, increased livestock mobility, livestock exclosures, grazing bans, rainwater harvesting infrastructure, education and skills development for primary stakeholders, social safety net programs to provide food and cash for villager implementation of interventions). Such programs should take advantage of low-cost options (such as self-scaling farmer-led restoration efforts) and be tailored to the local context which is based not only on who is involved, but also the farming system classification.

For example, priority policies and integrated landscape management initiatives for *mixed crop farming systems* should focus on supporting sustainable intensification of crop and livestock production (for example, securing upstream water supplies), reducing conflicts between farmers, and avoiding negative externalities of crop intensification (for example, agrochemical runoff that can pollute water bodies). Mixed crop farming systems are mostly prevalent in dry sub-humid zones and include farms that grow more than one crop type. All three case studies covered dry sub-humid zones and include mixed crop farming systems. These areas have some of the highest potential of Sub-Saharan African drylands for sustainable agricultural intensification and have a variety of land-use options whose trade-offs and opportunity costs must be reconciled. Some actions should be prioritized such as:

- Avoid negative externalities of intensification.
- Set up institutions for integrated land-use planning (that document and take into account existing rights including access to common pool resources and riparian areas) and mechanisms for conflict resolution.

- Support sustainable crop and livestock intensification that explores opportunities to take advantage of the landscape structure (mosaic of natural and managed ecosystems) to enhance biological control, pest management, pollination, or other ecosystem services, and that safeguards upstream water supplies and reduces downstream negative externalities on other water and natural resource users.

Agro-pastoral systems, as seen in the Ethiopia case study, combine pastoral or livestock livelihood systems based on animal production with crop production and exist widely across arid and semi-arid zones in Sub-Saharan Africa. Some of the biggest issues facing agro-pastoral systems that could undermine resilience-building efforts include farmer-herder conflicts and land tenure and rights. Policy recommendations and integrated landscape management initiatives must go beyond sustainable land management and soil and water conservation to reconcile stakeholder conflicts and provide better access to land resources. Priority recommendations should cover reducing conflicts, diversifying income sources, and improving resilience. For example, successful policies in agro-pastoral systems include:

- Establish corridors for livestock movement to protect farmers' crops and trees, designate grazing and water access areas to ensure resilience of pastoral systems, and set up dispute resolution mechanisms.
- Establish corridors for wildlife to access water and biomass during droughts.
- Support regeneration of dry forests and woodlands through assisted natural regeneration of trees and increase density of trees on farms through farmer-managed natural regeneration.
- Safeguard dry season grazing reserves (for example, wetlands, dry forests, and woodlands) and encourage planned grazing management.
- Develop water infrastructure that is aligned with sustainable forage management.
- Establish rewards or payments for biodiversity conservation, wildlife corridors and, in steep areas, watershed protection.
- Strengthen land tenure security and rights to manage trees on farms, which can become a catalyst to invest in and scale up the proposed actions above.

Finally, *pastoral systems*, which are mostly located in arid and semi-arid zones and include livestock livelihood systems focused on meat, milk, other animal products, and non-animal products, can benefit from integrated landscape management. In Sub-Saharan Africa, there are at least 25 million pastoralists dependent on livestock as their primary income source. Policy recommendations should also focus on resilience and livelihoods diversification, but some specific recommendations include:

- Support regeneration of pastoral landscapes through assisted natural regeneration of trees and shrubs with the help of exclosures and community-based natural resource management institutions (including customary pastoral institutions).

- Safeguard dry season grazing reserves (including wetlands) and access to and use of natural resources.
- Support pastoral mobility.
- Develop water infrastructure that is aligned with forage availability and grazing patterns to avoid risk of degradation.
- Encourage grazing management that improves soil cover, increases water infiltration and retention, and improves plant diversity and biomass.
- Establish rewards or payments for biodiversity conservation.
- Devolve management rights for pasturelands and water supplies as well as for wildlife, which can provide a strong local incentive for improved land, water, and natural resources management.

Conclusions

This book aimed to explore how integrated landscape management can reduce vulnerability and improve resilience of populations living in the drylands of Sub-Saharan Africa. It first defined landscapes and integrated landscape management, as well as a set of principles that could be used to define and develop landscape approaches. The book concluded by providing a variety of policy recommendations based on a study of integrated landscape approaches across the world, a stakeholder analysis, an analysis of unique economic and ecological benefits associated with integrated landscape approaches, and an in-depth look at three case studies of drylands in Sub-Saharan Africa.

Overall there is good evidence supporting how integrated landscape approaches that scale up improved land and water management practices can enhance the resilience of the most vulnerable herders and farmers and become the entry point for extending integrated landscape management across Africa's drylands. Increased investment in integrated landscape management programs, which support coordination and long-term collaboration among different groups of land managers and stakeholders within dryland landscapes, can enhance and safeguard these restoration efforts, lower risks related to water shortages and land degradation, diversify income sources, support sustainable intensification, and reduce conflicts.

While these recommendations are not comprehensive, they can serve to guide the international community including government officials and development cooperation experts in designing and implementing sustainable development programs to take advantage of the potential contributions of integrated landscape management to the restoration of resilience in the drylands of Sub-Saharan Africa.

References

Africa Livelihoods Networks Limited. 2013. "Mount Kenya East Pilot Project for Natural Resources Management Project Completion Report, February 2013." Government of Kenya, Ministry of Water and Irrigation and International Fund for Agriculture Development (IFAD), Nairobi, Kenya (accessed on February 28, 2014), http://www.mkepp.or.ke/index.php/other-reports/viewdownload/9-other-reports/21-mkepp-project-completion-report-2013.

African Union and NEPAD. 2009. "Sustainable Land and Water Management, The CAADP Pillar Framework, 'Tool' for Use by Countries in Mainstreaming and Upscaling of Sustainable Land and Water Management in Africa's Agriculture and Rural Development Agenda" (accessed on August 8, 2014), http://www.nepad-caadp.net/pdf/CAADP%20Pillar%201%20Framework.pdf

Albaugh, Janine M., Peter J. Dye, and John S. King. 2013. "Eucalyptus and Water Use in South Africa." *International Journal of Forestry Research*. http://www.hindawi.com/journals/ijfr/2013/852540/abs/.

Ammann, W. J., et al. 2013. "Background Document: The Economics of Desertification, Land Degradation and Drought: Methodologies and Analysis for Decision-Making." 2nd Scientific Conference United Nations Conference to Combat Desertification: Economic Assessment of Desertification, Sustainable Land Management and Resilience of Arid, Semi-Arid and Dry Sub-Humid Areas, April 9–12, Bonn, Germany.

Anderies, J. M., M. A. Janssen, and B. H. Walker. 2002. "Grazing Management, Resilience, and the Dynamics of a Fire-Driven Rangeland System." *Ecosystems* 5: 23–44.

Andersson, C., A. Mekonnen, and J. Stage. 2009. "Impacts of the Productive Safety Net Program in Ethiopia on livestock and Tree Holdings of Rural Households." Environment for Development. Discussion Paper Series, March 2009, EfD DP 09–05.

Bach, H., T. J. Clausen, D. T. Trang, L. Emerton, T. Facon, T. Hofer, K. Lazarus, C. Muziol, A. Noble, P. Schill, A. Sisouvanh, C. Wensley, and L. Whiting. 2011. "From Local Watershed Management to Integrated River Basin Management at National and Transboundary Levels." Mekong River Commission, Vientiane, Lao.

Barron, J., and S. Noel. 2011. "Valuing Soft Components in Agricultural Water Management Interventions in Meso-Scale Watersheds: A Review and Synthesis." *Water Alternatives* 42 (2): 145–54.

Belay, K.T., A. van Rompaey, J. Poesen, S. van Bruyssel, J. Deckers, and K.Amare. 2014. "Spatial analysis of land cover changes in Eastern Tigray (Ethiopia) from 1965 – 2007: are there signs of a forest transition?" In Land Degradation & Development. Published in Wiley online (wileyonlinelibrary.com) DOI: 10.1002/ldr.2275

Behnke, Roy, and Carol Kerven. 2011. "Replacing Pastoralism with Irrigated Agriculture in the Awash Valley, North-Eastern Ethiopia: Counting the Costs." Paper presented at the International Conference on Future of Pastoralism, March 21–23, 2011, organized by the Future Agricultures Consortium at the Institute of Development Studies, University of Sussex and the Feinstein International Center of Tufts University.

Berhane, G., J. Hoddinott, N. Kumar, and A. S. Taffesse. 2011. "The Impact of Ethiopia's Productive Safety Nets and Household Asset Building Programme: 2006–2010." International Food Policy Research Institute, Washington, DC.

Bewket, W. 2009. *Community-Based Rehabilitation of Degraded Lands: An Effective Response to Climate Change in Ethiopia*. World Food Programme. Addis Ababa, Ethiopia.

Birch, J., A. C. Newton, C. Alvarez Aquino, E. Cantarello, C. Echeverría, T. Kitzberger, I. Schiappaccasse, and N. Tejedor Garavito. 2010. "Cost-Effectiveness of Dryland Forest Restoration Evaluated by Spatial Analysis of Ecosystem Services." *Proceedings of the National Academy of Sciences* USA 107 (50): 21925–30.

Byrne, S., P. Fendrich, P. Arnold, and A. M. Acosta. 2011. "Four Case Studies on the Experience of SDC and Its Partners in Supporting Socially Inclusive Local Governance." Institute of Development Studies (accessed on February 28, 2014), http://www.sdc-decentralization.net/en/Home/Practical_Learning/Learning_Projects/Social_Inclusion.

Cadman, M., C. Petersen, A. Driver, N. Sekhran, K. Maze, and S. Munzhedzi. 2010. *Biodiversity for Development: South Africa's Landscape Approach to Conserving Biodiversity and Promoting Ecosystem Resilience*. Pretoria: South African National Biodiversity Institute.

Cameron, Edward. 2011. "From Vulnerability to Resilience: Farmer-Managed Natural Regeneration (FMNR) in Niger." Inside Stories on Climate Compatible Development. Climate and Development Knowledge Network (CDKN).

Capital Strategies Kenya Limited. 2012. "Mount Kenya East Pilot Project for Natural Resources Management (MKEPP–NRM) Impact Assessment Study Final Report, June 2012." Government of Kenya and International Fund for Agriculture Development (IFAD), Nairobi, Kenya (accessed on February 28, 2014) http://www.mkepp.or.ke/index.php/other-reports/viewdownload/9-other-reports/22-mkepp-impact-assessment-report-2012.

Cohen, M. J., M. Rocchigiani, and J. L. Garrett. 2008. "Empowering Communities through Food-Based Programmes: Ethiopia Case Study." Discussion Paper. World Food Programme.

Conservation South Africa. 2011. "Green Choice, Serving Nature's Bounty." http://biodiversityadvisor.sanbi.org/wp-content/uploads/2014/07/GreenChoice_brochure-2011_low-res.pdf.

Conway, Declan, E. Lisa, and F. Schipper. 2011. "Adaptation to Climate Change in Africa: Challenges and Opportunities Identified from Ethiopia." *Global Environmental Change* 21: 227–37.

Darghouth, S., C. Ward, G. Gambarelli, E. Styger, and J. Roux. 2008. *Watershed Management Approaches, Policies, and Operations: Lessons for Scaling Up*. Washington, DC: World Bank.

Deutsche Gesellschaft für Internationale Zusammenarbeit (GIZ). 2011. "Land Use Planning, Concept, Tools and Applications." GIZ. Eschborn, Germany (accessed on

August 8, 2014) http://www2.gtz.de/dokumente/bib-2011/giz2011-0041en-land-use-planning.pdf.

———. 2014. "Sustainable Land Management." Retrieved on February 20, 2014 from http://www.giz.de/en/worldwide/18912.html.

Dewees, P. A. 2013. "Bouncing Back: Forests, Trees and Resilient Households." Working Paper prepared for the International Conference on Forests for Food Security and Nutrition, Rome, May 13–15 (accessed on February 28, 2014) http://www.fao.org/forestry/37153-04fd46a0e4c02c12fb4631039a29b0c41.pdf.

Dosskey, M., G. Wells, G. Bentrup, and D. Wallace. 2012. "Enhancing Ecosystem Services: Designing for Multifunctionality." Soil and Water Conservation Society. *Journal of Soil and Water Conservation* 67 (2): 37A–41A (accessed on February 28, 2014) http://nac.unl.edu/documents/research/publications/2012DosskeyBentrupetalJSWC2012_enhanced.pdf.

Doyle, M., and C. A. Drew. 2008. *Large-Scale Ecosystem Restoration: Five Cases Studies from the United States*. Washington, DC: Island Press.

EcoAgriculture Partners. 2012a. *Reported Impacts of 23 Integrated Landscape Initiatives*. Washington, DC: EcoAgriculture Partners.

———. 2012b. "Restoration for Climate Resilience in the Tigray Region." EcoAgriculture. Retrieved on February 3, 2014 from http://blog.ecoagriculture.org/2012/11/30/meret/.

———. 2013a. *WFP Promotes Resilience in Chronic Food Insecure Areas of Ethiopia*. Washington, DC: EcoAgriculture Partners.

———. 2013b. "Defining Integrated Landscape Management for Policy Makers." EcoAgriculture Policy Focus No. 10, October 2013 (accessed on February 28, 2014) http://www.ecoagriculture.org/documents/files/doc_547.pdf.

Estrada-Carmona, Natalia, Abigail K. Hart, Fabrice A. J. DeClerck, Celia A. Harvey, and Jeffrey C. Milder. 2014. "Integrated Landscape Management for Agriculture, Rural Livelihoods, and Ecosystem Conservation: An Assessment of Experience from Latin America and the Caribbean." *Landscape and Urban Planning* 129 (September): 1–11.

Evans, A. E. V., M. Giordano, and T. Clayton, eds. 2012. "Investing in Agricultural Water Management to Benefit Smallholder Farmers in Ethiopia." AgWater Solutions, Project country synthesis report, International Water Management Institute (IWMI), Colombo, Sri Lanka.

Falkenmark, M., and J. Rockström. 2006. "The New Blue and Green Water Paradigm: Breaking New Ground for Water Resources Planning and Management." *Journal of Water Resources Planning and Management* (May–June): 120–33.

Flintan, F., R. Behnke, and C. Neely., 2013. "Natural Resource Management in the Drylands in the Horn of Africa." Brief prepared by a Technical Consortium hosted by CGIAR in partnership with the FAO Investment Centre, Technical Consortium Brief 1, International Livestock Research Institute, Nairobi.

Food and Agriculture Organization of the United Nations (FAO) 2000. "Conflict and Natural Resource Management." FAO.

———. 2011. *Payment for Ecosystem Services and Food Security*. Rome, Italy: FAO (accessed on February 28, 2014) http://www.fao.org/docrep/014/i2100e/i2100e.PDF.

————. 2014. "Sustainable Land Management" (accessed on July 22, 2014), http://www.fao.org/nr/land/sustainableland-management/en/.

Forest Trends, The Katoomba Group, and UNEP. 2008. "Payments for Ecosystem Services. Getting Started: A Primer." Forest Trends and The Katoomba Group.

Gebre Michael, Y., and A. Waters-Bayer. 2007. "Trees Are Our Backbone: Integrating Environment and Local Development in Tigray Region of Ethiopia." IIED Issue Paper no. 145, July.

Gebrewihot, T., and A. van der Veen. 2013. "Assessing the Evidence of Climate Variability in the Northern Part of Ethiopia." *Journal of Development and Agricultural Economics* 5 (3): 104–19.

German, L., H. Mansoor, G. Alemu, W. Mazengia, T. Amede, and A. Stroud.

2012. "Participatory Integrated Watershed Management: Evolution of Concepts and Methods." Working Paper 11, African Highlands Initiative.

Gichohi, H. W. 2003. "Direct Payments as a Mechanism for Conserving Important Wildlife Corridor Links between Nairobi National Park and Its Wider Ecosystem: The Wildlife Conservation Lease Program." World Parks Congress: Sustainable Finance Stream September 2003, Durban, South Africa Applications Session Learning from concrete successes of sustainably financing protected areas Workshop 9 Conservation Incentive Agreements (accessed on February 28, 2014) http://conservationfinance.org/guide/WPC/WPC_documents/Apps_09_Gichohi_v2.pdf.

Gikonyo, J. M., and C. M. Kiura. 2012. "Upper Tana Catchment Natural Resource Management Project (UTaNRMP)." Strategic Environmental Assessment Draft Report, August 2012. Government of Kenya (GOK) and International Fund for Agriculture Development (IFAD).

Gilligan, D. O., J. Hoddinott, and A. S. Taffesse. 2008. "The Impact of Ethiopia's Productive Safety Net Programme and Its Linkages." International Food Policy Research Institute Discussion Paper 00839.

Giordano, M., and T. Shah. 2014. "From IWRM Back to Integrated Water Resources Management." *International Journal of Water Resources Development*: 1–13 (accessed on February 28, 2014) http://www.tandfonline.com/doi/pdf/10.1080/07900627.2013.851521.

Global Environment Facility. 2004. Project of the Government of Kenya, IFAD/UNEP GEF Project, GEF ID 1848, Mount Kenya East Pilot Project for Natural Resources Management, Project Brief Report (accessed on February 28, 2014). http://www.thegef.org/gef/sites/thegef.org/files/repository/KenyaMKEPP.pdf.

————. 2014. "The Great Green Wall Initiative." Retrieved on July 23, 2014 from http://www.thegef.org/gef/great-green-wall.

Global Mechanism of the UNCCD, FAO. 2009. "Policy and Financing for Sustainable Land Management in Sub-Saharan Africa: Lessons and Guidance for Action Version 1.0." TerrAfrica.

Global Risk Forum. 2013. "The Economics of Desertification, Land Degradation and Drought: Methodologies and Analysis for Decision-Making." UNCCD Background document for the 2nd Scientific Conference UNCCD, Bonn, Germany.

Goldstein, Joshua H., Giorgio Caldarone, Chris Colvin, T. Ka'eo Duarte, Driss Ennaanay, Kalani Fronda, Neil Hannahs, Emily McKenzie, Guillermo Mendoza, Kapu Smith, Stacie Wolny, Ulalia Woodside, and Gretchen C. Daily, *TEEBcase*. "The Natural Capital Project,

Kamehameha Schools, and InVEST: Integrating Ecosystem Services into Land-Use Planning in Hawai`i." 2010.

Goldstein, Joshua H., Giorgio Caldarone, Thomas Kaeo Duarte, Driss Ennaanay, Neil Hannahs, Guillermo Mendoza, Stephen Polasky, Stacie Wolny, and Gretchen C. Daily. 2012. "Integrating Ecosystem-Service Tradeoffs into Land-Use Decisions." *PNAS* 109 (19): 7565–70.

Government of Ethiopia. 2012. "Ethiopian Government Portal, Regional States, State Tigray" (accessed on February 20, 2014) Retrieved on February 20, 2014 from http://www.ethiopia.gov.et/statetigray.

Government of India National Rainfed Area Authority (NRAA). 2011. "Common Guidelines for Watershed Development Projects—2008." Revised edition 2011, Government of India, New Delhi.

Government of India Planning Commission. 2012. "Twelfth Five Year Plan (2012–2017): Faster, More Inclusive and Sustainable Growth." Government of India. Retrieved on January 2, 2013 from http://planningcommission.gov.in/plans/planrel/12thplan/pdf/vol_1.pdf.

Gray, E., and A. Srinidhi. 2013. "Watershed Development in India: Economic Valuation and Adaptation Considerations." Working Paper, World Resources Institute, Washington, DC.

Grimble, R., M.-K. Chan, J. Aglionby, and J. Quan. 1995. "Trees and Trade-Offs: A Stakeholder Approach to Natural Resource Management." Gatekeeper Series No. 52, International Institute for Environment and Development, Sustainable Agriculture and Rural Livelihoods Programme.

Hagazi, N., and G. Hailemariam. 2012. "Effect of Watershed Management Efforts in Climate Change Adaptation and Livelihood Improvement: The Case of Abreha Atsebeha Peasant Association, Tigray, Ethiopia." Tigray Agricultural Research Institute.

Haglund E. et al. 2010. "Assessing the Impacts of Farmer-Managed Natural Regeneration in the Sahel: A Case Study of the Maradi Region, Niger." Cited in C. Pye-Smith (2013) "The Quiet Revolution: How Niger's Farmers are Regreening the Croplands of the Sahel," World Agroforestry Center (ICRAF), Trees for Change no. 12, World Agroforestry Center, Nairobi.

Haglund, E., J. Ndjeunga, L. Snook, and D. Pasternak. 2011. "Dryland Tree Management for Improved Household Livelihoods: Farmer Managed Natural Regeneration in Niger." *Journal of Environmental Management* 92 (7): 1696–705. Elsevier Ltd.

Haregeweyn, N., A. Berhe, A. Tsunekawa, M. Tsubo, and D. T. Meshesha. 2012. "Integrated Watershed Management as an Effective Approach to Curb Land Degradation: A Case Study of the Enabered Watershed in Northern Ethiopia." *Environmental Management* 50: 1219–33.

Hassane, A., P. Martin, and C. Reij. 2000. "Water Harvesting, Land Rehabilitation and Household Food Security in Niger: IFAD's Soil and Water Conservation Project in Illéla District." VU University Amsterdam and International Fund for Agricultural Development, Rome.

Headey, D., M. Dereje, J. Ricker-Gilbert, A. Josephson, and A. S. Taffesse. 2013. "Land Constraints and Agricultural Intensification in Ethiopia: A Village-Level Analysis of High-Potential Areas." Ethiopia Strategy Support Program Working Paper 58, Ethiopian Development Research Institute and the International Food Policy Research Institute.

Hunger, Nutrition, Climate Justice. 2013. "Scaling Up an Integrated Watershed Management Approach through Social Protection Programmes in Ethiopia: The MERET and PSNP schemes." A New Dialogue: Putting People at the Heart of Global Development (accessed on February 28, 2014) http://www.mrfcj.org/pdf/case-studies/2013-04-16-Ethiopia-MERET.pdf.

International Fund for Agriculture Development (IFAD). 2012a. "Upper Tana Catchment Natural Resource Management Project (UTaNRMP)." Project Design Report, Main Report, IFAD, East and Southern Africa Division, Project Management Department, February 2012 (accessed on February 28, 2014) http://www.ifad.org/operations/projects/design/105/kenya.pdf.

———. 2012b. "Supervision Report of the International Fund for Agricultural Development." Country: Republic of Kenya; Project: Mount Kenya East Pilot Project for Natural Resources Management; IFAD Loan: 599–KE, June, IFAD, Nairobi, Kenya (accessed on February 28, 2014) http://operations.ifad.org/documents/654016/68a57ef1-82b3-4c53-92e7-a18adb64f2b4.

International Union for Conservation of Nature, Eastern and Southern Africa Regional Office (IUCN-ESARO), 2010. "Drylands Situation Analysis" (accessed on November 13, 2013) https://cmsdata.iucn.org/downloads/iucn_esaro_drylands_situation_analysis.pdf.

———. 2012. *Supporting Pastoral Livelihoods: A Global Perspective on Minimum Standards and Good Practices*. Second Edition, March 2012. Nairobi, Kenya: IUCN-ESARO.

Joshi, P. K., A. K. Jha, S. P. Wani, L. Joshi, and R. L. Shiyani. 2005. "Meta-Analysis to Assess the Impact of Watershed Program and People's Participation." Comprehensive Assessment Research Report 8. Comprehensive Assessment Secretariat, Colombo, Sri Lanka.

Kale, G., V. L. Manekar, and P. D. Porey. 2012. "Watershed Development Project Justification by Economic Evaluation: A Case Study of Kachhighati Watershed in Aurangabad District, Maharashtra." *ISH Journal of Hydraulic Engineering* 18 (2): 101–11.

Karl, M. 2000. "Monitoring and Evaluating Stakeholder Participation in Agriculture and Rural Development Projects: A Literature Review." Food and Agriculture Organization of the United Nations.

Kassa, K.T. 2013. "Detection and Analysis of Land Use and Land Cover Changes in Tigray, North Ethiopia." PhD Katholieke Universiteit Leuven.

Kellert, S. A., J. N. Mehta, S. A. Ebbin, and L. L. Lichtenfeld. 2000. "Community Natural Resource Management: Promise, Rhetoric, and Reality." *Society and Natural Resources* 13: 705–15. Retrieved on November 8, 2014 from http://afrpw.org/wp-content/uploads/2011/12/Community-Natural-Resource-Management.Lichtenfeld.pdf .

Kerr. 2002. "Watershed Development Projects in India: An Evaluation." International Food Policy Research Institute, Washington, DC.

Kissinger, G., A. Brasser, and L. Gross. 2013. "Reducing Risk: Landscape Approaches to Sustainable Sourcing." Synthesis Report, Washington, DC. Landscapes for People, Food and Nature Initiative (accessed on February 28, 2014) http://landscapes.ecoagriculture.org/documents/files/reducing_risk_synthesis_report.pdf.

Kremen, A., and A. Miles. 2012. "Ecosystem Services in Biologically Diversified versus Conventional Farming Systems: Benefits, Externalities, and Trade-Offs." *Ecology and Society* 17 (4): 40 (accessed on February 28, 2014) http://www.ecologyandsociety.org/vol17/iss4/art40/.

Kristjanson, P., M. Radeny, D. Nkedianye, R. Kruska, R. Reid., H. Gichohi, F. Atieno, and R. Sanford. 2002. "Valuing Alternative Land-Use Options in the Kitengela Wildlife Dispersal Area of Kenya." ILRI Impact Assessment Series No 10. A joint ILRI (International Livestock Research Institute)/ACC (African Conservation Centre) report, Nairobi, Kenya.

Kumasi, T. C., and K. Asenso-Okyere. 2011. "Responding to Land Degradation in the Highlands of Tigray, Northern Ethiopia." International Food Policy Research Institute. Discussion Paper 01142.

Lakew D., V. Carucci, A. Wendem-Ageliehu, and Y. Abebe, eds. 2005. "Community-Based Participatory Watershed Development: A Guideline." Federal Democratic Republic of Ethiopia, Ministry of Agriculture and Rural Development, Addis Ababa, Ethiopia (accessed on February 28, 2014) ftp://bsesrv214.bse.vt.edu/Dillaha/ReferenceMaterial/CommunityBasedParticipatoryWatershedDevelopmentAGuidelinePart1.pdf.

Larwanou, M., M. Abdoulaye, and C. Reij. 2006. "Etude la regeneration naturelle assistée dans la Région de Zinder (Niger)." International Resources Group/USAID.

Learning Initiative. 2012. "Learning Initiative 2012: Making Rangelands Secure; Protecting Livestock Mobility Routes: Lessons Learned" (accessed on February 28, 2014) http://www.celep.info/wp-content/uploads/2013/09/2012-Protecting-Livestock-Mobility-Lessons-Learned.pdf.

Liniger, H. P., R. Mekdaschi Studer, C. Hauert, and M. Gurtner. 2011. "Sustainable Land Management in Practice—Guidelines and Best Practices for Sub-Saharan Africa." TerrAfrica, World Overview of Conservation Approaches and Technologies (WOCAT) and Food and Agriculture Organization of the United Nations (FAO) (accessed on February 28, 2014) http://www.wocat.net/fileadmin/user_upload/documents/Books/SLM_in_Practice_E_Final_low.pdf.

Lipper, L., T. Sakuyama, R. Stringer, and D. Zilberman. 2009. *Payment for Environmental Services in Agricultural Landscapes: Economic Policies and Poverty Reduction in Developing Countries.* Rome, Italy: Food and Agriculture Organization of the United Nations (FAO).

Mackedon, J. 2012. "Rehabilitating China's Loess Plateau. Scaling Up in Agriculture, Rural Development and Nutrition." Focus 19, Brief 5, June, International Food Policy Research Institute.

Marchal, J. Y. 1979. "L'espace des techniciens et celui des paysans: historique d'un perimeter anti-érosif en Haute-Volta." *Maîtrise de l'espace agraire et développement en Afrique tropicale*, 245–52. PARIS: ORSTOM.

Mayunga, J. S. 2007. "Understanding and Applying the Concept of Community Disaster Resilience: A Capital-Based Approach." Draft working paper prepared for social vulnerability and resilience building, Munich, Germany.

Meikle, A. 2010. "Africa Climate Change Resilience Alliance (ACCRA): Ethiopia." Country level literature review, ACCRA.

Meire, E., A. Frankl, A. de Wulf, A. M. Haile, J. Deckers, and J.Nyssen. 2012. "Land Use and Cover Dynamics in Africa since the Nineteenth Century: Warped Terrestrial Photographs of North Africa." In *Regional Environmental Change*, published online, November.

Milder, J. C., A. K. Hart, P. Dobie, and J. Minai. 2014. "Integrated Landscape Initiatives for African Agriculture, Development, and Conservation: A Region-Wide Assessment." *World Development* 54: 68–80.

Millennium Ecosystem Assessment. 2005. *Ecosystems and Human Well-Being: Synthesis.* Washington, DC: Island Press.

Ministry of Agriculture and Rural Development 2011. "Revised Project Implementation Manual for the Sustainable Land Management Program." The Federal Democratic Republic of Ethiopia, MoARD.

Mogaka, Hezron, Samuel Gichere, Richard Davis, and Rafik Hirji. 2005. "Climate Variability and Water Resources Degradation in Kenya: Improving Water Resources Development and Management." World Bank Working Paper 69, World Bank, Washington, DC.

Mogoi, J., J. Tanui, W. Mazengia, and C. Lyamchai. 2009. "Role of Collective Action and Policy Options for Fostering Participation in Natural Resource Management." Paper presented during the 2nd World Congress on Agroforestry, August 24–28, Nairobi, Kenya.

Mortimore, M., M. Tiffen, and F. Gichuki. 1993. "Sustainable Growth in Machakos." *ELEIA Newsletter* 9 (4) (accessed on February 28, 2014) http://www.agriculturesnetwork.org/magazines/global/strong-case-for-diversity/sustainable-growth-in-machakos/at_download/article_pdf.

Muchena, F. N., D. D. Onduru, and J. H. Kauffman. 2011. "Analysis of Financial Mechanisms for Green Water Credits in the Upper Tana, Kenya." Green Water Credits Report 17, Series Editors, W. R. S. Critchley, E. M. Mollee, ISRIC World Soil Information, Wageningen, Netherlands (accessed on February 28, 2014) http://www.greenwatercredits.net/sites/default/files/documents/isric_gwc_report17.pdf.

Nedessa, B. 2013. "Experiences of the MERET Project of the Ministry of Agriculture in Integrated Watershed Management in Ethiopia." Powerpoint presentation at the NBDC Regional Stakeholder Dialogue, Bahir Day, July 23–24.

Nedessa, B., and S. Wickrema. 2010. "Disaster Risk Reduction: Experience from the MERET Project in Ethiopia." In *Revolution: From food aid to food assistance,* edited by S. W. Omamo, U. Gentilini, and S. Sandström, 139–56. Rome: World Food Programme.

Nega, F., E. Mathijs, J. Deckers, M. Haile, J. Nyssen, and E. Tollens. 2008. "Rural Poverty Dynamics and Impact of Intervention Programs upon Chronic and Transitory Poverty in Northern Ethiopia." Paper presented at WIDER conference on Frontiers of Poverty Analysis, United Nations University. Available at http://www.afdb.org/fileadmin/uploads/afdb/Documents/Knowledge/30753357-EN-133-FREDU-RURAL-POVERTY-DYNAMICS.PDF

Neufeldt, H., et al., 2013. "Beyond Climate-Smart Agriculture: Toward Safe Operating Spaces for Global Food Systems." *Agriculture and Food Security* 2: 12. Retrieved on November 8, 2014 from http://www.agricultureandfoodsecurity.com/content/pdf/2048-7010-2-12.pdf.

Newton, A. C., R. F. del Castillo, C. Echeverria, D. Geneletti, M. Gonzalez-Espinosa, L. R. Malizia, A. C. Prmoli, J. M. R. Benayas, C. Smith-Ramirez, and G. Williams-Linera. 2012. "Forest Landscape Eestoration in the Drylands of Latin America." *Ecology and Society* 17 (1): 21.

Nsouli, S. M. 2000. "Capacity Building in Africa: The Role of International Financial Institutions." *Finance and Development* 37 (4): 34–7. International Monetary Fund.

Nyssen, J., N. Munro, M. Haile, J. Poesen, K. Deeschemaeker, N. Haregeweyn, J. Moeyersons, G. Govers, and J. Deckers. 2007. "Understanding the Environmental Changes in Tigray: A Photographic Record Over 30 Years." Tigray Livelihood Papers No. 3, VLIR—Mekelle University IUC Programme and Zala-Daget Project, 82p.

OECD. 2013. "Providing Agri-Environmental Public Good through Collective Action." Joint Working Party on Agriculture and the Environment (accessed on February 28, 2014) http://search.oecd.org/officialdocuments/publicdisplaydocumentpdf/?cote=C OM/TAD/CA/ENV/EPOC%282012%2911/FINAL&docLanguage=En.

Onduru, D. D., and F. N. Muchena. 2011. "Cost Benefit Analysis of Land Management Options in the Upper Tana, Kenya." Green Water Credits Report 15, Series Editors, W.R.S. Critchley, E.M. Mollee, ISRIC World Soil Information, Wageningen, Netherlands (accessed on February 28, 2014) http://www.isric.org/sites/default/files/isric_gwc_ report15.pdf.

Organisation for Economic Co-operation and Development. 2013. "Providing Agri-Environmental Public Goods through Collective Action." OECD Joint Working Party on Agriculture and the Environment.

Pagiola, S. 2008. "Payments for Environmental Services in Costa Rica." *Journal of Ecological Economics* 65: 712–24.

Pagiola, S., and A. Arcenas. 2013. "Regional Integrated Silvopastoral Ecosystem Management Project—Costa Rica, Colombia and Nicaragua." The Economics of Ecosystems and Biodiversity.

Pagiola, S., A. R. Rios, and A. Arcenas. 2008. "Can the Poor Participate in Payments for Environmental Services? Lessons from the Silvopastoral Project in Nicaragua." *Journal of Environment and Development Economics* 13: 299–325.

Palanisami, K., D. S. Kumar, S. P. Wani, and M. Giordano. 2009. "Evaluation of Watershed Development Programmes in India Using Economic Surplus Method." *Agricultural Economics Research Review* 22 (July–December): 197–207.

Pender, J., F. Place, and S. Ehui, eds. 2006. *Strategies for Sustainable Land Management in the East African Highlands.* Washington, DC: International Food Policy Research Institute.

Place, F., and D. Garrity. 2014. "Tree-Based Systems to Increase Resilience in Drylands." Background paper to the Economics of Drylands.

Porras, I., M. Grieg-Gran, and G. Meijerink. 2007. "Farmers' Adoption of Soil and Water Conservation: Potential Role of Payments for Watershed Services." Green Water Credits Report 5, ISRIC—World Soil Information, Wageningen, The Netherlands (accessed on February 28, 2014) http://www.greenwatercredits.net/sites/default/files/ documents/gwc_report_5.pdf.

Pye-Smith, C. 2013. "The Quiet Revolution: How Niger's Farmers Are Re-Greening the Parklands of the Sahel." ICRAF Trees for Change no. 12, World Agroforestry Centre, Nairobi.

Reij, C. 2013. "Climate-Smart Agriculture, Food Security and Water in Africa's Drylands: Lessons from Experience" (Presentation at World Resources Institute).

Reij, C., G. Tappan, and A. Belemvire. 2005. "Changing Land Management Practices and Vegetation on the Central Plateau of Burkina Faso (1968–2002)." *Journal of Arid Environments* 63: 642–59.

Reij, C., G. Tappan, and M. Smale. 2009. "Agroenvironmental Transformation in the Sahel: Another Kind of Green Revolution." IFPRI Discussion Paper 00914, Paper prepared for Millions Fed: proven successes in agricultural development.

Riley, B., A. Ferguson, S. Ashine, C. Torres, and S. Asnake. 2009. "Mid-term Evaluation of the Ethiopia Country Programme 10430.0 (2007–2011): Final Report." World Food Programme, Office of Evaluation.

Rinaudo, T., and A. Admasu. 2010. "Agricultural Development Recommendations Tigray Region." World Vision Ethiopia Task Force Report.

Ruhweza, A., B. Biryahwaho, and C. Kalanzi. 2008. "An Inventory of PES Schemes in Uganda." Retrieved on November 15, 2013 from http://www.katoombagroup.org/regions/africa/documents/2008_Uganda_Inventory.pdf.

Sayer, J., T. Sunderland, J. Ghazoul, J. L. Pfund, D. Sheil, E. Mcijaard, M. Venter, A. Klintuni Boedhihartono, M. Day, C. Garcia, C. van Oosten, and L. E. Buck. 2013. "Ten Principles for a Landscape Approach to Reconciling Agriculture, Conservation, and Other Competing Land Uses." *Proceedings of the National Academy of Science* 110 (21): 8349–56 (accessed on February 28, 2014) http://www.pnas.org/content/early/2013/05/14/1210595110.full.pdf+html.

Schmidt, E., and F. Tadess. 2012. "Household and Plot Level Impact of Sustainable Land and Watershed Management Practices in the Blue Nile." Ethiopian Development Research Institute and International Food Policy Research Institute, ESSP Research Note 18.

Sendzimir, J., C. P. Reij, and P. Magnuszewski. 2011. "Rebuilding Resilience in the Sahel: Regreening in the Maradi and Zinder Regions of Niger." *Ecology and Society* 16 (3): 1.

Shames, S., A. Wekesa, and E. Wachiye. 2012. "Case Study: Western Kenya Smallholder Agriculture Carbon Finance Project: Vi Agroforestry." CGIAR Research Program on Climate Change, Agriculture and Food Security (CCAFS) Institutional innovations in African smallholder carbon projects, EcoAgriculture Partners and Vi Agroforestry, June (accessed on February 28, 2014) http://cgspace.cgiar.org/bitstream/handle/10568/21222/AfricanAgCarbon-CaseStudy-ViAgroforestry.pdf?sequence=6.

Shames, S., M. H. Clarvis, and G. Kissinger. 2013. *Financing Strategies for Integrated Landscape Management: Implications for Climate Policy. Policy Focus*. Washington, DC: EcoAgriculture.

Sharma, S. 2012. "An Update on the World Bank's Experimentation with Soil Carbon." Promise of Kenya Agricultural Carbon Project Remains Elusive. Published October 4 (accessed on February 28, 2014) http://www.iatp.org/documents/an-update-on-the-world-bank%E2%80%99s-experimentation-with-soil-carbon.

Sreedevi, T.K., S. P. Wani, R. Sudi, M. S. Patel, T. Jayesh, S. N. Singh, and S. Tushar. 2006. "On-site and Off-site Impact of Watershed Development: A Case Study of Rajasamadhiyala, Gujarat, India." Global Theme on Agroecosystems Report No. 20, Patancheru 502 324, Andhra Pradesh, India: International Crops Research Institute for the Semi-Arid Tropics.

Stevens, C., R. Winterbottom, J. Springer, and K. Reytar. 2014. *Securing Rights, Combating Climate Change: How Strengthening Community Forest Rights Mitigates Climate Change*. Washington, DC: World Resources Institute.

Sukhdev, P., H. Wittmer, and D. Miller. 2014. "The Economics of Ecosystems and Biodiversity (TEEB): Challenges and Responses." In D. Helm and C. Hepburn (eds), *Nature in the Balance: The Economics of Biodiversity*. Oxford: Oxford University Press.

Sutter, P., T. Frankenberger, J. Downen, M. Greeley, and M. Mueller. 2012. "Ethiopia: MERET Impact Evaluation." World Food Programme. TANGO International, Institute of Development Studies, and Ethiopian Economics Association.

Talberth, J., E. Gray, E. Branosky, and T. Garrner. 2012. "Insights from the Field: Forests for Water." WRI Issue Brief. Southern Forests for the Future Incentives Series. Issue Brief 9 (accessed on February 28, 2014) http://pdf.wri.org/insights_from_the_field_forests_for_water.pdf.

TEEB. 2009. "The Economics of Ecosystems and Biodiversity for National and International Policy-Makers—Summary: Responding to the Value of Nature."

Thalmeinerova, D., and S. Downey. 2013. "Cease-Fire on IWRM" (accessed on February 28, 2014) http://wle.cgiar.org/blogs/2013/05/14/cease-fire-on-iwrm/.

Tiffen, M., M. Mortimore, and F. Gichuki. 1994. "More People, Less Erosion." Environmental Recovery in Kenya, Overseas Development Institute, London (accessed on February 28, 2014) http://www.odi.org.uk/sites/odi.org.uk/files/odi-assets/publications-opinion-files/4600.pdf.

Tittonel, P. A. 2013. "Farming Systems Ecology. Towards ecological intensification of world agriculture." Inaugural lecture upon taking up the position of Chair in Farming Systems Ecology at Wageningen University on May 16, Wageningen University, Wageningen, The Netherlands.

USAID. 2012. "USAID's Legacy in Agriculture: Integrating Natural Resources Management into Agricultural Practices and Livelihoods." Working paper prepared by WRI.

———. 2014. "An Overview of USAID's Credit Guarantees." Retrieved on February 26, 2014 from http://www.usaid.gov/sites/default/files/documents/2151/DCAOnePager.pdf.

van Noordwijk, M., S. Bruijnzeel, D. Ellison, D. Sheil, C. Morris, D. Sands, V. Gutierrez, J. Cohen, C. A. Sullivan, B. Verbist, D. Murdiyarso, D. Gaveau, and B. Muys. 2015. "Ecological Rainfall Infrastructure: Investment in Trees for Sustainable Development." ASB Policy Brief 47. SB Partnership for the Tropical Forest Margins, Nairobi.

van Steenbergen, F., A. Tuinhof, and L. Knoop. 2011. "Transforming Lives Transforming Landscapes." The Business of Sustainable Water Buffer Management, 3R Water Secretariat Wageningen, The Netherlands (accessed on February 28, 2014) http://www.hydrology.nl/images/docs/ihp/nl/2011.08_Transforming_Landscapes.pdf.

Verhagen, J., T. Vellinga, F. Neijenhuis, T. Jarvis, L. Jackson, P. Caron, P. Torquebiau, L. Lipper, E. Fernandes, R. E. M. Entsuah, and S. Vermeulen. 2014. "Climate Smart Agriculture." Wageningen University and Research Center, UC Davis, Centre de coopération internationale en recherche agronomique pour le Développement, Food and Agriculture Organization of the United Nations, World Bank, Council for Scientific and Industrial Research, Ghana, CGIAR, and Climate Change, Agriculture and Food Security.

Wagkari, S. 2009. "Customary Pastoral Institution Study." Report for the World Initiative for Sustainable Pastoralism, Save the Children USA, http://www.iucn.org/wisp/resources/.

Winterbottom, R., C. Reij, D. Garrity, J. Glover, and D. Hellums. 2013. "Improving Land and Water Management." Working Paper, World Resources Institute, Washington, DC (accessed on February 28, 2014) http://www.wri.org/sites/default/files/improving_land_and_water_management_0.pdf.

World Bank Institute. 2010. *Rehabilitating a Degraded Watershed. A Case Study from China's Loess Plateau*. Washington, DC: World Bank Group.

World Bank. 2006a. *Ethiopia—Sustainable Land Management Program Project*. Washington, DC: World Bank

———. 2006b. *Managing Water Resources to Maximize Sustainable Growth: A Country Water Resources Assistance Strategy for Ethiopia*. Washington, DC: World Bank.

———. 2007. "Restoring China's Loess Plateau" (accessed on July 22, 2014) from http://www.worldbank.org/en/news/feature/2007/03/15/restoring-chinas-loess-plateau.

———. 2008. "Forest Source Book. Practical Guidance for Sustaining Forests in Development Cooperation." Washington, DC (accessed on February 28, 2014) http://siteresources.worldbank.org/EXTFORSOUBOOK/Resources/completeforestsource-bookapril2008.pdf.

———. 2009a. "Convenient Solutions to an Inconvenient Truth: Ecosystem-based Approaches to Climate Change." June 2009 (accessed on February 28, 2014) http://siteresources.worldbank.org/ENVIRONMENT/Resources/ESW_EcosystemBasedApp.pdf.

———. 2009b. "Report No: ICR00001067. Implementation Completion and Results Report (TF–50723) for a Northern Savanna Biodiversity Project, Ghana." World Bank, Washington, DC.

———. 2011. "Report No: ICR00001787, Implementation Completion and Results Report (TF–53855) for a Sahel Integrated Lowland Ecosystem Management Project (SILEM), Burkina Faso." World Bank, Washington, DC.

———. 2012. "Ethiopia Climate Project Receives Africa's First Forestry Carbon Credits under CDM." Retrieved on February 28, 2014 from http://www.worldbank.org/en/news/feature/2012/10/09/ethiopia-climate-project-receives-africa-s-first-forestry-carbon-credits.

———. 2013. Concept Note on the Economics of Drylands (mimeo).

———. 2014a. "Ethiopian Experience in Sustainable Land Management." Draft—detailed annex to concept note.

———. 2014b. "Kenyans Earn First Ever Carbon Credits from Sustainable Farming." January 21, 2014 (accessed on February 28, 2014) http://www.worldbank.org/en/news/press-release/2014/01/21/kenyans-earn-first-ever-carbon-credits-from-sustainable-farming.

World Bank, Rapid Social Response, and Global Facility for Disaster Reduction and Recovery. 2013. *Ethiopia's Productive Safety Net Program (PSNP): Integrating Disaster and Climate Risk Management*. Washington, DC: World Bank.

World Economic Forum. 2010. "Realizing a New Vision for Agriculture: A Roadmap for Stakeholders" (accessed on February 28, 2014) http://www3.weforum.org/docs/WEF_IP_NVA_Roadmap_Report.pdf.

World Food Programme. 1997. "Interim Evaluation of Project Ethiopia 2488 (Exp. 3): Rehabilitation and Development of Rural Lands and Infrastructure." World Food Programme.

———. 2013. "Annual Report 2012." World Food Programme Ethiopia.

World Food Programme Ethiopia. 2005. "Report on the Cost-Benefit Analysis and Impact Evaluation of Soil and Water Conservation and Forestry Measures." Managing

Environmental Resources to Enable Transitions to More Sustainable Livelihoods, WFP, Addis Ababa, Ethiopia.

World Resources Institute (WRI) in collaboration with United Nations Development Programme, United Nations Environment Programme, and World Bank. 2005. *World Resources 2005: The Wealth of the Poor-Managing Ecosystems to Fight Poverty.* Washington, DC: WRI (accessed on February 28, 2014) http://pdf.wri.org/wrr05_lores.pdf.

World Resources Institute (WRI); Department of Resource Surveys and Remote Sensing, Ministry of Environment and Natural Resources, Kenya; Central Bureau of Statistics, Ministry of Planning and National Development, Kenya; and International Livestock Research Institute. 2007. *Nature's Benefits in Kenya, An Atlas of Ecosystems and Human Well-Being.* Washington, DC and Nairobi: World Resources Institute (accessed on February 28, 2014) http://www.wri.org/sites/default/files/kenya_atlas_full-text_150.pdf.

World Resources Institute in collaboration with United Nations Development Programme, United Nations Environment Programme, and World Bank. 2008. *World Resources 2008: Roots of Resilience.* Washington, DC: WRI.

Yamba, B., and M. N. Sambo. 2012. "La régénération naturelle assistée et la sécurité alimentaire des ménages de 5 terroirs villageois de départements de Kantché et Mirriah (région de Zinder, Niger)." Report for the International Fund for Agricultural Development. Etude FIDA 1246-VU University Amsterdam.

Yitbarek, T. W., S. Belliethathan, and M. Fetene. 2010. "A Cost-Benefit Analysis of Watershed Rehabilitation: A Case Study in Farta Woreda, South Gondar, Ethiopia." *Ecological Restoration* 28 (1): 46–55.

Zeleke, G., M. Kassie, J. Pender, and M. Yesuf. 2006. "Stakeholder Analysis for Sustainable Land Management (SLM) in Ethiopia: Assessment of Opportunities, Strategic Constraints, Information Needs, and Knowledge Gaps." Environmental Economics Policy Forum for Ethiopia and International Food and Policy Research Institute.

www.ingramcontent.com/pod-product-compliance
Lightning Source LLC
Chambersburg PA
CBHW080423270326
41929CB00018B/3138